# Functional Analysis and Summability

# Functional Analysis and Summability

**P.N. Natarajan**

CRC Press
Taylor & Francis Group
Boca Raton London New York

CRC Press is an imprint of the
Taylor & Francis Group, an **informa** business
A CHAPMAN & HALL BOOK

First edition published 2021
by CRC Press
6000 Broken Sound Parkway NW, Suite 300, Boca Raton, FL 33487-2742

and by CRC Press
4 Park Square, Milton Park, Abingdon, Oxon OX14 4RN


CRC Press is an imprint of Taylor & Francis Group, an Informa business



***Library of Congress Cataloging-in-Publication Data***

Names: Natarajan, P. N., author.
Title: Functional analysis and summability / P.N. Natarajan.
Description: First edition. | Boca Raton, FL : CRC Press, 2021. |
Includes bibliographical references and index.
Identifiers: LCCN 2020017697 (print) | LCCN 2020017698 (ebook) |
ISBN 9780367544492 (hardback) | ISBN 9781003089360 (ebook)
Subjects: LCSH: Functional analysis. | Summability theory.
Classification: LCC QA320 .N38 2021 (print) | LCC QA320 (ebook) |
DDC 515/.7--dc23
LC record available at https://lccn.loc.gov/2020017697
LC ebook record available at https://lccn.loc.gov/2020017698

**Visit the Taylor & Francis Web site at**
**http://www.taylorandfrancis.com**

**and the CRC Press Web site at**
**http://www.crcpress.com**

ISBN 13: 978-0-367-54449-2 (hbk)

Dedicated to my paternal grandfather, Vidyaratna Dr. P.S. Subrahmanya Sastri (1890-1978), a great linguist, on his 130th birth anniversary.

# *Contents*

# *About the Author*

**P.N. Natarajan** (Pinnangudi Narayanasubramanian Natarajan) is a former professor and head of the Department of mathematics at Ramakrishna Mission Vivekananda College, Chennai. He received his Ph.D. in Analysis from the University of Madras in 1980. An active researcher throughout his career, Prof. Natarajan has more than 120 research papers to his credit, published in reputable journals like *Proceedings of the American Mathematical Society*, *Journal of the London Mathematical Society*, *Indagationes Mathematicae*, *Annales Mathematiques Blaise Pascal*, *Commentationes Mathematicae (Prace Matematyczne)*, *p-adic Numbers*, *Ultrametric Analysis and Applications*, and *Journal of the Indian Mathematical Society*. He has so far authored five books, published by Springer-Verlag, John Wiley, and Taylor and Francis. He has also contributed a chapter to each of three edited volumes published by Springer-Verlag, Birkhauser-Verlag, and Taylor and Francis. His research interests include summability theory and functional analysis, both classical and ultrametric. Prof. Natarajan was honoured with the Dr. Radhakrishnan Award for the Best Teacher in mathematics for the year 1990–1991 by the Government of Tamil Nadu. In addition to visiting several important institutes in Canada, France, Holland, and Greece on invitation, Prof. Natarajan has taken part in several international conferences and has chaired sessions.

# About the Author

# Index of Symbols

# Preface

There are excellent books on functional analysis but only a few of them are really introductory. The master work *Théorie Des Opérations Linéaires* (1932) of S. Banach stands out as the first and foremost among the several excellent books on functional analysis. Study of divergent series is the foundation of summability theory. Again there are excellent books on summability theory, the outstanding among them being *Divergent Series* (1949) of G.H. Hardy. Most of these are terse. My own experience in teaching these topics to several batches of post-graduate students pinpointed to me that they needed a more gentle introduction of these topics. I have written the present book with this point in mind. While writing the book, I have been influenced by many authors; of whom, special mention has to be made of G. Bachman and L. Narici [1], C. Goffman and G. Pedrick [7], G.H. Hardy [8], I.J. Maddox [13], and G.F. Simmons [24]. Though the book is primarily meant for students of mathematics with basic knowledge of real and complex analysis, it will also be useful to students of physics and engineering to get some flavour of functional analysis and summability. The topic on hand has many utilities in applied mathematics too. A physicist or engineer, who works on Fourier Series, Fourier transforms, or analytic continuation, can also find the book very useful for his/her research.

In this book, spread over nine chapters, a lot of examples have been provided at appropriate places to aid student understanding, and exercises have been given at needed places.

We now briefly describe the contents of the various chapters of the book.

In Chapter 1, we introduce several basic concepts in functional analysis such as linear space, inner product space, normed linear space, metric space, completeness of a metric space, and Banach space with numerous examples.

Chapter 2 is devoted to a study of factor space, linear transformation, continuity of a linear transformation, linear functionals, bounded linear functionals, and then a conjugate space. Further, we prove some results on finite-dimensional spaces and then study convexity.

In Chapter 3, we prove the Hahn–Banach theorem for real-normed linear spaces and complex-normed linear spaces. Further, we study some consequences of the Hahn–Banach theorem.

In Chapter 4, we study second conjugate spaces, reflexivity, and proceed to find the conjugate spaces of $\ell_1$, $\ell_p$, $p > 1$, $c_0$ and $C[a, b]$.

In Chapter 5, we introduce several types of convergence. We then prove the Banach–Steinhaus theorem (or the uniform boundedness principle). Some consequences and applications of the Banach–Steinhaus theorem are then discussed.

In Chapter 6, we study the bounded inverse theorem, closed graph theorem, interior mapping principle, and the open mapping theorem.

In Chapter 7, we introduce Hilbert spaces, orthogonality, orthogonal complements, and orthonormal basis, and establish the Riesz representation theorem for linear functionals on a Hilbert space. With Chapter 7, we conclude the functional analysis component of the present book.

With Chapter 8, we commence the summability component of the book. In Chapter 8, we prove the famous Silverman–Toeplitz theorem on regularity of matrix transformations. We give several examples of regular methods. We then proceed to prove Schur's theorem. We also record Hahn's version of Schur's theorem.

In Chapter 9, we establish the Steinhaus theorem, again a famous result in summability theory. We also prove several Steinhaus-type theorems in the remaining part of the chapter.

In the summability component of the book, we give instances where functional analytic tools like the uniform boundedness principle are used to prove very important theorems.

The author thanks Mr. Boopal Ethirajan for expertly typing the manuscript. Constructive suggestions for the improvement of the book are always welcome.

**P.N. Natarajan**

*Chennai, India*

*December 15, 2019*

# Chapter 1

# *Some Basic Concepts in Functional Analysis*

## 1.1 Linear space, inner product space, and normed linear space: Examples

We now recall a few definitions and concepts from linear algebra that are needed in the sequel.

**Definition 1.** *A linear space or vector space over a field $K$ is a quadruple $(X, K, +, \cdot)$, where $X$ is a non-empty set, $+$ is a mapping $(x, y) \to x + y$ of $X \times X \to X$, $\cdot$ is a mapping $(\alpha, x) \to \alpha \cdot x$ of $K \times X \to X$, and the following conditions are fulfilled:*

*(i)* $x + (y + z) = (x + y) + z$, *for all* $x, y, z \in X$;

*(ii)* $x + y = y + x$, *for all* $x, y \in X$;

*(iii) there exists* $\theta \in X$ *such that* $x + \theta = x$, *for all* $x \in X$;

*(iv) for each* $x \in X$, *there exists an element denoted by* $-x \in X$ *such that* $x + (-x) = \theta$;

*(v)* $(\alpha + \beta) \cdot x = \alpha \cdot x + \beta \cdot x$, *for all* $x \in X$, $\alpha, \beta \in K$;

*(vi)* $\alpha \cdot (x + y) = \alpha \cdot x + \alpha \cdot y$, *for all* $x, y \in X$, $\alpha \in K$;

*(vii)* $\alpha.(\beta.x) = (\alpha\beta).x$, *for all* $x \in X$, $\alpha, \beta \in K$;

*and*

*(viii)* $1.x = x$, *for all* $x \in X$, *where 1 is the identity element of* $K$.

We usually speak of the linear space $X$ in place of $(X, K, +, .)$. When we want to mention the field $K$, we speak of the linear space $(X, K)$.

A linear space over $\mathbb{R}$ (the field of real numbers) is called a "real linear space," whereas a linear space over $\mathbb{C}$ (the field of complex numbers) is called a "complex linear space." The elements of $X$ are called "vectors or points," whereas the elements of $K$ are called "scalars." $K$ itself is called the "scalar field." Throughout the present book, $K = \mathbb{R}$ or $\mathbb{C}$.

We note that a complex linear space $(X, \mathbb{C}, +, .)$ is also a real linear space $(X, \mathbb{R}, +, .)$, which is obtained by restricting the mapping $(\alpha, x) \to \alpha.x$ of $\mathbb{C} \times X \to X$ into $\mathbb{R} \times X$. $(X, \mathbb{R}, +, .)$ is called the real linear space associated with the complex linear space $(X, \mathbb{C}, +, .)$.

Let $X$ be a linear space. It is easy to check that the elements $\theta$, $-x$, mentioned in (iii), (iv) of Definition 1 are unique. $\theta$ is called the zero of the linear space $X$. The following results are easily verified.

$$(-1).x = -x, \text{ for all } x \in X; \tag{1.1}$$

$$0.x = \theta, \text{ for all } x \in X; \tag{1.2}$$

and

$$\alpha.\theta = \theta, \text{ for all } x \in K. \tag{1.3}$$

For the sake of convenience, hereafter, we use the symbol "0" to denote the zero of $K$ and the zero of $X$ too.

**Definition 2.** *For* $x_1, x_2, \ldots, x_n \in X$ *and* $\alpha_1, \alpha_2, \ldots, \alpha_n \in K$, *the element* $\alpha_1 x_1 + \alpha_2 x_2 + \cdots + \alpha_n x_n \in X$ *is called a "linear combination of* $x_1, x_2, \ldots, x_n$.*"*

**Definition 3.** *A subset $A$ of $X$ is said to "span" $X$, if every element of $X$ can be expressed as a linear combination of elements of $A$.*

**Definition 4.** *$x_1, x_2, \ldots, x_n \in X$ are said to be "linearly dependent," if there exist $\alpha_1, \alpha_2, \ldots, \alpha_n \in K$, not all zero, such that*

$$\alpha_1 x_1 + \alpha_2 x_2 + \cdots + \alpha_n x_n = 0.$$

*A non-empty subset $A$ of $X$ is said to be "linearly independent" if there exists no finite subset of $A$, whose elements are linearly dependent.*

**Definition 5.** *A linearly independent subset $A$ of $X$ is said to be a "basis" of $X$ if every element of $X$ can be expressed uniquely as a linear combination of elements of $A$.*

**Definition 6.** *$X$ is said to be "finite-dimensional" if $X = \{0\}$ or $X$ has a finite basis. Otherwise, $X$ is said to be "infinite-dimensional." One can prove that if $X$ is finite-dimensional and $X \neq \{0\}$, then all the bases of $X$ are finite and have the same number of elements. In such a case, the number of elements in a basis is called the "dimension of $X$." We shall define the dimension of $\{0\}$ as 0.*

We now give a few examples of linear spaces.

**Example 1.** *Let $\mathbb{R}^n$ be the set of all ordered $n$-tuples of real numbers. For $x = (x_1, x_2, \ldots, x_n)$, $y = (y_1, y_2, \ldots, y_n) \in \mathbb{R}^n$, $x_i, y_i \in \mathbb{R}$, $i = 1, 2, \ldots, n$, define*

$$x + y = (x_1 + y_1, x_2 + y_2, \ldots, x_n + y_n),$$

$$\alpha x = (\alpha x_1, \alpha x_2, \ldots, \alpha x_n), \alpha \in \mathbb{R}.$$

*With respect to the operations defined above, one can verify that $\mathbb{R}^n$ is a linear space over $\mathbb{R}$. It is called the "$n$-dimensional real space." Similarly, we can define the "$n$-dimensional complex space.".*

**Example 2.** *Consider the set $C[a,b]$ of all real-valued (complex-valued) continuous functions defined on $[a,b]$ with the usual operations of addition of functions and multiplication of functions by numbers. Then $C[a,b]$ is a linear space over $\mathbb{R}$ $(\mathbb{C})$.*

**Example 3.** *Consider the set $\ell_2$ of all infinite sequences $\{x_n\}$ of real (complex) numbers such that*

$$\sum_{i=1}^{\infty} |x_i|^2 < \infty \tag{1.4}$$

*with the following operations of addition and scalar multiplication:*

$$\{x_n\} + \{y_n\} = \{x_n + y_n\};$$
$$\alpha(\{x_n\}) = \{\alpha x_n\}, \alpha \in \mathbb{R}(\mathbb{C}).$$

*$\ell_2$ is then a linear space over $\mathbb{R}$ $(\mathbb{C})$.*

**Example 4.** *The set $c$ of all convergent sequences, the set $c_0$ of all null sequences, i.e., all sequences converging to zero, the set $\ell_\infty$ of all bounded sequences, and the set $\mathbb{R}^\infty$ $(\mathbb{C}^\infty)$ of all sequences of real (complex) numbers, with respect to the operations defined in Example 3, are linear spaces over $\mathbb{R}$ $(\mathbb{C})$.*

**Definition 7.** *Let $S$ be a non-empty subset of a linear space $X$. If $S$ itself is a linear space with respect to the operations of addition and scalar multiplication defined in $X$, then $S$ is called a "linear subspace" or simply a "subspace" of $X$.*

It is easy to prove that $S$ is a subspace of $X$ if and only if

$$\alpha x + \beta y \in S, \text{ whenever } x, y \in S \text{ and } \alpha, \beta \in K.$$

We note that $\{0\}$ is a subspace of $X$ and $X$ itself if a subspace of $X$. These two subspaces are known as "improper or trivial subspaces of $X$." Any other subspace of $X$ is called a "proper subspace of $X$."

We now provide some examples of subspaces.

**Example 5.** *Let $X$ be a linear space over $K$ and $x \in X$. Then*

$$\{\lambda x / \lambda \in K\}$$

*is a subspace of $X$.*

**Example 6.** *The set $P[a, b]$ of all polynomials defined on $[a, b]$ is a subspace of $C[a, b]$.*

**Example 7.** *Each of the linear spaces $\ell_2, c_0, c, \ell_\infty, \mathbb{R}^\infty$ ($\mathbb{C}^\infty$) is a subspace of the succeeding one.*

**Definition 8.** *Let $X$ be a linear space and $U$ be a subset of $X$. Then the intersection of all linear subspaces of $X$ containing $U$ is called the "linear hull" of $U$. It is also called the subspace of $X$ generated by $U$ and is denoted by $[U]$.*

We note that the linear hull of $U$ is a subspace of $X$ containing $U$ and it is the smallest subspace of $X$ containing $U$. We also note that the linear hull of $U$ is the set of all finite linear combinations of elements of $U$.

**Definition 9.** *Let $X$ be a linear space over $K$. A function $f : X \to K$ is called a "functional."*

**Definition 10.** *A functional $f$ is said to be "additive" if*

$$f(x + y) = f(x) + f(y), \text{for } x, y \in X.$$

*$f$ is said to be "homogeneous" if*

$$f(\alpha x) = \alpha f(x), \text{for } x \in X, \alpha \in K.$$

*An additive, homogeneous functional $f$ is called a "linear functional."*

**Definition 11.** *A function $f$ defined on a complex linear space $X$ is said to "conjugate homogeneous" if*

$$f(\alpha x) = \overline{\alpha} f(x), \text{for } x \in X, \alpha \in \mathbb{C},$$

*$\overline{\alpha}$ being the complex conjugate of $\alpha$. A conjugate homogeneous, additive, functional is called a "conjugate, linear functional."*

We give a few examples of linear functionals.

**Example 8.** *Let* $x = (x_1, x_2, \ldots, x_n) \in \mathbb{R}^n$ *and* $a = (a_1, a_2, \ldots, a_n) \in \mathbb{R}^n$ *be a fixed element. Define*

$$f : \mathbb{R}^n \to \mathbb{R}$$

*by*

$$f(x) = \sum_{i=1}^{n} a_i x_i.$$

*Then* $f$ *is a linear functional on* $\mathbb{R}^n$*.*

**Example 9.** *As in the previous example, one can prove that* $f$ *defined by*

$$f(x) = \sum_{i=1}^{n} a_i \overline{x}_i$$

*is a conjugate linear functional on* $\mathbb{C}^n$*.*

We now introduce "inner product spaces."

**Definition 12.** *Given a linear space* $X$ *over* $\mathbb{C}$ *(or* $\mathbb{R}$*), by an inner product on* $X$*, we mean a mapping*

$$<,>: X \times X \to \mathbb{C}$$

$$(x, y) \to < x, y >$$

*such that*

*(i)* $< x, y >= \overline{< y, x >}$, *for* $x, y \in X$, $-$ *denoting conjugation in* $\mathbb{C}$*;*

*(ii)* $< x + y, z >=< x, z > + < y, z >$, *for* $x, y, z \in X$*;*

*(iii)* $< \alpha x, y >= \alpha < x, y >$, *for* $x, y \in X$, $\alpha \in \mathbb{C}$*;*

*(iv)* $< x, x >\geq 0$*;*

*and*

*(v)* $< x, x >= 0$ *if and only if* $x = 0$*.*

*The linear space $X$, with the inner product $<,>$, is called an "inner product space," denoted by $(X, <,>)$. For convenience, usually we drop $<,>$ and denote the space by $X$.*

We note the following:

$$\begin{aligned} < x, \alpha y > &= \overline{< \alpha y, x >} = \overline{\alpha < y, x >} \\ &= \overline{\alpha} \overline{< y, x >} \\ &= \overline{\alpha} < x, y >, \text{for } x, y \in X, \alpha \in \mathbb{C} \end{aligned}$$

and

$$\begin{aligned} < x, y + z > &= \overline{< y + z, x >} \\ &= \overline{< y, x > + < z, x >} \\ &= \overline{< y, x >} + \overline{< z, x >} \\ &= < x, y > + < x, z >, \text{for } x, y, z \in X. \end{aligned}$$

**Theorem 1** (Cauchy–Schwarz inequality). *In an inner product space $X$,*

$$| < x, y > | \leq < x, x >^{\frac{1}{2}} < y, y >^{\frac{1}{2}}, \textit{for } x, y \in X.$$

*Proof.* Let $A = < x, x >$, $C = < y, y >$, $B = | < x, y > |$. Now, for $\lambda \in K$,

$$\begin{aligned} < x + \lambda y, x + \lambda y > &= < x, x > + < x, \lambda y > + < \lambda y, x > + < \lambda y, \lambda y > \\ &= < x, x > + \overline{\lambda} < x, y > + \lambda < y, x > + \lambda \overline{\lambda} < y, y > \\ &= < x, x > + 2RP\overline{\lambda} < x, y > + |\lambda|^2 < y, y > \\ &= A + 2RP\overline{\lambda} < x, y > + |\lambda|^2 C. \end{aligned}$$

Since $< x + \lambda y, x + \lambda y > \geq 0$,

$$A + 2RP\overline{\lambda} < x, y > + |\lambda|^2 C \geq 0.$$

Let $< x, y > = re^{i\theta}$ and $\lambda = \mu e^{i\theta}$, $r, \mu \in \mathbb{R}$.

Then

$$\begin{aligned} \overline{\lambda} < x, y > &= \mu e^{-i\theta} . re^{i\theta} \\ &= \mu r, \mu r \in \mathbb{R}. \end{aligned}$$

Thus,

$$|\lambda|^2 C + 2\mu r + A \geq 0.$$

Since $|\lambda| = \mu$, $| < x, y > | = r$, we have

$$C\mu^2 + 2| < x, y > |\mu + A \geq 0,$$
$$i.e., C\mu^2 + 2B\mu + A \geq 0,$$

from which it follows that

$$4B^2 \leq 4AC,$$
$$i.e., B^2 \leq AC,$$
$$i.e., | < x, y > |^2 \leq < x, x >< y, y >,$$
$$i.e., | < x, y > | \leq < x, x >^{\frac{1}{2}} < y, y >^{\frac{1}{2}},$$

completing the proof. □

Some examples of inner product spaces are given below.

**Example 10.** *Let* $X = \mathbb{C}^n$, *the set of all* $n$*-tuples of complex numbers. For* $x = (x_1, x_2, \ldots, x_n)$, $y = (y_1, y_2, \ldots, y_n) \in X$, *we define*

$$< x, y >= \sum_{i=1}^{n} x_i \overline{y}_i.$$

*We can show that* $<, >$ *is an inner product on* $\mathbb{C}^n$. $\mathbb{C}^n$, *with the above inner product, is called the "complex Euclidean* $n$*-space." With* $\mathbb{C}$ *replaced by* $\mathbb{R}$, *we get the "real Euclidean* $n$*-space."*

**Example 11.** *Let* $X = C[a, b]$, *the set of all complex-valued continuous functions defined on* $[a, b]$. *For* $f, g \in X$, *define*

$$< f, g >= \int_a^b f(x)\overline{g(x)}dx.$$

*One can prove that* $<, >$ *is an inner product on* $X$.

**Example 12.** *Let* $X = \ell_2$. *For* $x = \{x_n\}$, $y = \{y_n\} \in X$, *define*

$$< x, y >= \sum_{i=1}^{\infty} x_i \overline{y}_i.$$

*In this context, we note that* $\sum_{i=1}^{\infty} x_i \overline{y}_i$ *converges in view of (1.4) and Theorem 1. We can now prove that* $<,>$ *is an inner product on* $X$.

**Definition 13.** *A normed linear space* $X$ *is a linear space over* $\mathbb{C}$ *(or* $\mathbb{R}$*) in which there is a mapping*

$$\|.\| : X \to \mathbb{R}$$

*such that*

*(i)* $\|x\| \geq 0$ *and* $\|x\| = 0$ *if and only if* $x = 0$;

*(ii)* $\|\alpha x\| = |\alpha| \|x\|$;

*and*

*(iii)* $\|x + y\| \leq \|x\| + \|y\|$ *(triangle inequality),*

*where* $x, y \in X$ *and* $\alpha \in K$. *Thus the normed linear space is the pair* $(X, \|.\|)$. *For convenience, we denote the space by* $X$ *itself.*

**Theorem 2.** *Every inner product space is a normed linear space.*

*Proof.* Let $X$ be an inner product space with inner product $<,>$. Define $\|.\| : X \to \mathbb{R}$ by

$$\|x\| =< x, x >^{\frac{1}{2}}, \text{for } x \in X.$$

We first note that $\|x\| \geq 0$ and $\|x\| = 0$ if and only if $x = 0$. Now, for $x \in X$ and $\alpha \in K$,

$$\begin{aligned} \|\alpha x\|^2 &=< \alpha x, \alpha x > \\ &= \alpha \overline{\alpha} < x, x > \\ &= |\alpha|^2 \|x\|^2 \end{aligned}$$

and consequently,

$$\|\alpha x\| = |\alpha|\|x\|.$$

Again, for $x, y \in X$,

$$\begin{aligned}
\|x+y\|^2 &= < x+y, x+y > \\
&= < x,x > + < x,y > + < y,x > + < y,y > \\
&= \|x\|^2 + < x,y > + \overline{< x,y >} + \|y\|^2 \\
&= \|x\|^2 + 2RP < x,y > + \|y\|^2 \\
&\leq \|x\|^2 + 2\|x\|\|y\| + \|y\|^2, \text{ using Theorem 1} \\
&= (\|x\| + \|y\|)^2,
\end{aligned}$$

from which we have

$$\|x+y\| \leq \|x\| + \|y\|,$$

completing the proof. □

---

## 1.2 Metric space, examples, Banach space, and examples of Banach spaces

**Definition 14.** *By a metric on a non-empty set $X$, we mean a mapping*

$$\begin{aligned}
d : X \times X &\to \mathbb{R} \\
(x, y) &\to d(x, y)
\end{aligned}$$

*such that*

*(i)* $d(x, y) \geq 0$, $d(x, y) = 0$ *if and only if* $x = y$;

*(ii)* $d(x, y) = d(y, x)$;

*and*

*(iii)* $d(x, y) \leq d(x, z) + d(z, y)$ *(triangle inequality),*

*where* $x, y, z \in X$. *The non-empty set* $X$, *together with the metric* $d$, *is called a metric space, denoted by* $(X, d)$. *For convenience, we denote the metric space by* $X$ *itself as usual.*

**Theorem 3.** *Every normed linear space is a metric space.*

*Proof.* Let $X$ be a normed linear space, with norm $\|.\|$.
Define

$$d : X \times X \to \mathbb{R}$$

by

$$d(x, y) = \|x - y\|, \text{for } x, y \in X.$$

We first note that $d(x, y) \geq 0$ and

$$d(x, y) = 0$$

if and only if

$$\|x - y\| = 0,$$

$$i.e., x - y = 0,$$

$$i.e., x = y.$$

Also

$$\begin{aligned} d(x, y) &= \|x - y\| \\ &= \| - (y - x)\| \\ &= | - 1|\|y - x\| \\ &= \|y - x\| \\ &= d(y, x). \end{aligned}$$

Finally,

$$\begin{aligned} d(x, y) &= \|x - y\| \\ &= \|(x - z) + (z - y)\| \\ &\leq \|x - z\| + \|z - y\| \\ &= d(x, z) + d(z, y), \end{aligned}$$

for $x, y, z \in X$. Thus $d$ is a metric and so $X$ is a metric space. □

The reader can give some examples of normed linear spaces. We shall now provide a few examples of metric spaces.

**Example 13.** *Let $X$ be any non-empty set. Consider the function*

$$d : X \times X \to \mathbb{R}$$

*defined by*

$$d(x, y) = \begin{cases} 1, & \text{if } x \neq y; \\ 0, & \text{if } x = y, \end{cases}$$

*$x, y \in X$.*
*It is easy to prove that d is a metric on $X$. This metric is called the "trivial" metric on $X$. Any other metric on $X$ is said to be "non-trivial."*

**Example 14.** *The function $d : \mathbb{R} \times \mathbb{R} \to \mathbb{R}$ defined by*

$$d(x, y) = |x - y|, \text{for } x, y \in \mathbb{R}$$

*is a metric on $\mathbb{R}$.*

**Example 15.** *Let $X = \mathbb{R}^2$. For $x = (x_1, x_2)$, $y = (y_1, y_2) \in \mathbb{R}^2$, define*

*(i)* $d_1(x, y) = \{(x_1 - y_1)^2 + (x_2 - y_2)^2\}^{\frac{1}{2}}$;

*(ii)* $d_2(x, y) = |x_1 - y_1| + |x_2 - y_2|$;

*(iii)* $d_3(x, y) = \max(|x_1 - y_1|, |x_2 - y_2|)$.

The reader can check that $d_1, d_2, d_3$ are metrices on $\mathbb{R}^2$. These can be generalized for $\mathbb{R}^n$ too.

**Example 16.** *Let $X = C[a, b]$, the set of all complex-valued continuous functions defined on $[a, b]$. For $f, g \in X$, define*

$$d(f, g) = \max_{x \in X} |f(x) - g(x)|.$$

*Then $d$ is a metric on $X$.*

**Example 17.** *Let $(X_i, d_i)$, $i = 1, 2, \ldots, n$ be a finite collection of metric spaces. Consider the Cartesian product*

$$X = X_1 \times X_2 \times \cdots \times X_n.$$

*For $x = (x_1, x_2, \ldots, x_n)$, $y = (y_1, y_2, \ldots, y_n) \in X$, define*

*(i)* $d(x, y) = \sum_{i=1}^{n} d_i(x_i, y_i);$

*(ii)* $d(x, y) = \max_{1 \leq i \leq n} d_i(x_i, y_i);$

*(iii)* $d(x, y) = \left[ \sum_{i=1}^{n} \{d_i(x_i, y_i)\}^2 \right]^{\frac{1}{2}}.$

We can prove that $d$, in each case, is a metric on $X$.

**Definition 15.** *Let $(X, d)$ be a metric space and $S$ be a subset of $X$. If $S$ itself is a metric space with respect to the metric $d$, $S$ is called a "metric subspace" or simply a "subspace" of $X$.*

**Definition 16.** *Let $(X, d)$ be a metric space. Let $\{x_n\}$ be a sequence of points in $X$. We say that $\{x_n\}$ converges to $x \in X$, if for any $\epsilon > 0$, there exists a positive integer $N$ such that*

$$d(x_n, x) < \epsilon, \text{ for all } n \geq N.$$

*In such a case, we write*

$$\lim_{n \to \infty} x_n = x \text{ or } x_n \to x, n \to \infty.$$

*$x$ is called the limit of the sequence $\{x_n\}$. We note that whenever $x$ exists, it is unique.*

**Definition 17.** *Let $(X, d)$ be a metric space. A sequence $\{x_n\}$ in $X$ is called a "Cauchy sequence," if for any $\epsilon > 0$, there exists a positive integer $N$ such that*

$$d(x_m, x_n) < \epsilon, \text{ for all } m, n \geq N.$$

Note that every convergent sequence is a Cauchy sequence. The converse is not true. Consider $\mathbb{Q}$, the set of all rational numbers, with the metric $d$ defined by

$$d(x, y) = |x - y|, \text{ for } x, y \in \mathbb{Q}.$$

Let

$$x_n = \left(1 + \frac{1}{n}\right)^n, n = 1, 2, \ldots.$$

We know that $\{x_n\}$ converges to $e$. Since $\{x_n\}$ converges, it is a Cauchy sequence. However, its limit $e \notin \mathbb{Q}$.

**Definition 18.** *If every Cauchy sequence in a metric space $X$ converges to an element of $X$, $X$ is said to be "complete."*

**Theorem 4.** *Addition and scalar multiplication in a normed linear space are jointly continuous, i.e., if*

$$x_n \to x, y_n \to y, \alpha_n \to \alpha, n \to \infty,$$

*$x_n, y_n \in X$, $n = 1, 2, \ldots, x, y \in X$, $\alpha_n \in K$, $n = 1, 2, \ldots$, $\alpha \in K$, then*

*(i)* $x_n + y_n \to x + y, n \to \infty$

*and*

*(ii)* $\alpha_n x_n \to \alpha x, n \to \infty.$

*Proof.* (i) Given $\epsilon > 0$, we can find a positive integer $N$ such that, for all $n \geq N$,

$$\|x_n - x\| < \frac{\epsilon}{2}$$

and

$$\|y_n - y\| < \frac{\epsilon}{2}.$$

Now,

$$\begin{aligned}\|(x_n + y_n) - (x + y)\| &= \|(x_n - x) + (y_n - y)\| \\ &\leq \|x_n - x\| + \|y_n - y\| \\ &< \frac{\epsilon}{2} + \frac{\epsilon}{2} \\ &= \epsilon, \text{ for all } n \geq N,\end{aligned}$$

$$i.e., x_n + y_n \to x + y, n \to \infty.$$

(ii) Since $\{\alpha_n\}$ converges, it is bounded, i.e., there exists $M > 0$ such that

$$|\alpha_n| \leq M, \text{ for all } n = 1, 2, \ldots.$$

Now, given any $\epsilon > 0$, we can find a positive integer $N$ such that

$$|\alpha_n - \alpha| < \frac{\epsilon}{2\|x\|}$$

and

$$\|x_n - x\| < \frac{\epsilon}{2M},$$

for all $n \geq N$. Thus,

$$\begin{aligned}\|\alpha_n x_n - \alpha x\| &= \|\alpha_n(x_n - x) + (\alpha_n - \alpha)x\| \\ &\leq \|\alpha_n(x_n - x)\| + \|(\alpha_n - \alpha)x\| \\ &= |\alpha_n|\|x_n - x\| + |\alpha_n - \alpha|\|x\| \\ &< M\frac{\epsilon}{2M} + \frac{\epsilon}{2\|x\|}\|x\|, \text{ for all } n \geq N \\ &= \frac{\epsilon}{2} + \frac{\epsilon}{2} \\ &= \epsilon, \text{ for all } n \geq N,\end{aligned}$$

$$i.e., \alpha_n x_n \to \alpha x, n \to \infty,$$

completing the proof. □

**Theorem 5.** *Let $X$ be a normed linear space with $\|.\|$. Then*

$$\|.\| : X \to \mathbb{R}$$

*is a continuous mapping.*

*Proof.* First we observe that

$$|\|x\| - \|y\|| \leq \|x - y\|.$$

Let $x_n \to x$, $n \to \infty$ in $X$. Given $\epsilon > 0$, there exists a positive integer $N$ such that

$$\|x_n - x\| < \epsilon, \text{ for all } n \geq N.$$

Now,

$$\begin{aligned} |\|x_n\| - \|x\|| &\leq \|x_n - x\| \\ &< \epsilon, \text{ for all } n \geq N, \\ \text{i.e., } \|x_n\| &\to \|x\|, n \to \infty, \\ \text{i.e., } \|.\| & \text{ is a continuous mapping.} \end{aligned}$$

□

We have seen that the norm in a normed linear space induces a metric. If $X$ is complete with respect this metric, $X$ is known as a "Banach space." In other words, a Banach space is a complete normed linear space. A Banach space, in which the norm is induced by an inner product, is called a "Hilbert space."

We now provide a few examples of Banach spaces. Toward this end, we establish some important inequalities:

**Lemma 1.**
*(i)* $\displaystyle\sum_{i=1}^{n} |x_i y_i| \leq \left(\sum_{i=1}^{n} |x_i|^2\right)^{\frac{1}{2}} \left(\sum_{i=1}^{n} |y_i|^2\right)^{\frac{1}{2}}$ *(Cauchy's inequality);*

*(ii)* $\displaystyle\left(\sum_{i=1}^{n} |x_i + y_i|^2\right)^{\frac{1}{2}} \leq \left(\sum_{i=1}^{n} |x_i|^2\right)^{\frac{1}{2}} + \left(\sum_{i=1}^{n} |y_i|^2\right)^{\frac{1}{2}}$,
*where* $x_i, y_i \in \mathbb{C}$, $i = 1, 2, \ldots, n$.

*Proof.* (i) Let $a_i = \dfrac{|x_i|^2}{\sum\limits_{i=1}^{n} |x_i|^2}$, $b_i = \dfrac{|y_i|^2}{\sum\limits_{i=1}^{n} |y_i|^2}$, $i = 1, 2, \ldots, n$. We know that

$$(a_i b_i)^{\frac{1}{2}} \leq \frac{a_i + b_i}{2}, i = 1, 2, \ldots, n,$$

$$i.e., \frac{|x_i||y_i|}{\left(\sum\limits_{i=1}^{n} |x_i|^2\right)^{\frac{1}{2}} \left(\sum\limits_{i=1}^{n} |y_i|^2\right)^{\frac{1}{2}}} \leq \frac{1}{2} \left( \frac{|x_i|^2}{\sum\limits_{i=1}^{n} |x_i|^2} + \frac{|y_i|^2}{\sum\limits_{i=1}^{n} |y_i|^2} \right),$$

$i = 1, 2, \ldots, n$. Summing this for $i = 1, 2, \ldots, n$, we get

$$\frac{\sum\limits_{i=1}^{n} |x_i y_i|}{\left(\sum\limits_{i=1}^{n} |x_i|^2\right)^{\frac{1}{2}} \left(\sum\limits_{i=1}^{n} |y_i|^2\right)^{\frac{1}{2}}} \leq \frac{1}{2}(1 + 1)$$

$$= 1,$$

from which we get

$$\sum_{i=1}^{n} |x_i y_i| \leq \left(\sum_{i=1}^{n} |x_i|^2\right)^{\frac{1}{2}} \left(\sum_{i=1}^{n} |y_i|^2\right)^{\frac{1}{2}}.$$

(ii) $|x_i + y_i|^2 \leq |x_i + y_i|(|x_i| + |y_i|)$, $i = 1, 2, \ldots, n$.

Thus,

$$
\begin{aligned}
\sum_{i=1}^{n}|x_i+y_i|^2 &\le \sum_{i=1}^{n}|x_i+y_i|(|x_i|+|y_i|)\\
&= \sum_{i=1}^{n}|x_i||x_i+y_i| + \sum_{i=1}^{n}|y_i||x_i+y_i|\\
&\le \left(\sum_{i=1}^{n}|x_i|^2\right)^{\frac{1}{2}}\left(\sum_{i=1}^{n}|x_i+y_i|^2\right)^{\frac{1}{2}}\\
&\quad + \left(\sum_{i=1}^{n}|y_i|^2\right)^{\frac{1}{2}}\left(\sum_{i=1}^{n}|x_i+y_i|^2\right)^{\frac{1}{2}},\\
&\quad \text{in view of Cauchy's inequality (i)}\\
&= \left[\left(\sum_{i=1}^{n}|x_i|^2\right)^{\frac{1}{2}} + \left(\sum_{i=1}^{n}|y_i|^2\right)^{\frac{1}{2}}\right]\\
&\quad \times \left(\sum_{i=1}^{n}|x_i+y_i|^2\right)^{\frac{1}{2}}.
\end{aligned}
$$

If $\sum_{i=1}^{n}|x_i+y_i|^2 \neq 0$, then

$$
\left(\sum_{i=1}^{n}|x_i+y_i|^2\right)^{\frac{1}{2}} \le \left(\sum_{i=1}^{n}|x_i|^2\right)^{\frac{1}{2}} + \left(\sum_{i=1}^{n}|y_i|^2\right)^{\frac{1}{2}},
$$

noting that the above inequality trivially holds if $\sum_{i=1}^{n}|x_i+y_i|^2 = 0$, completing the proof. □

**Lemma 2.** *(i) If $a, b \ge 0$, then*

$$
a^{\frac{1}{p}}b^{\frac{1}{q}} \le \frac{a}{p} + \frac{b}{q},
$$

*where $1 < p < \infty$ and $\frac{1}{p} + \frac{1}{q} = 1$;*

*(ii)* $\sum_{i=1}^{n}|x_iy_i| \le \left(\sum_{i=1}^{n}|x_i|^p\right)^{\frac{1}{p}}\left(\sum_{i=1}^{n}|y_i|^q\right)^{\frac{1}{q}}$ *(Hölder's inequality);*

*(iii)* $\|x+y\|_p \le \|x\|_p + \|y\|_p$, $1 \le p < \infty$,

*where*

$$
\|x\|_p = \left(\sum_{i=1}^{n}|x_i|^p\right)^{\frac{1}{p}},
$$

$x = (x_1, x_2, \ldots, x_n)$, $y = (y_1, y_2, \ldots, y_n)$, $x_i, y_i \in \mathbb{C}$, $i = 1, 2, \ldots, n$.

*Proof.* (i) If $a$ or $b = 0$, the inequality trivially holds. Now, let $a, b \neq 0$. For $k \in (0,1)$, define

$$f(t) = k(t-1) - t^k + 1,$$

where $t \geq 1$. Note that $f(1) = 0$. If $t > 1$,

$$\begin{aligned} f'(t) &= k - kt^{k-1} \\ &= k\left(1 - \frac{1}{t^{1-k}}\right) \\ &> 0, \end{aligned}$$

so that

$$\begin{aligned} f(t) &\geq 0, \text{ for } t \geq 1, \\ i.e., t^k &\leq kt + (1-k). \end{aligned}$$

Case 1. Let $a \geq b$. Then $\frac{a}{b} \geq 1$ so that

$$\left(\frac{a}{b}\right)^k \leq k\left(\frac{a}{b}\right) + (1-k).$$

Put $k = \frac{1}{p}$ so that $k \in (0,1)$. Thus,

$$\begin{aligned} \left(\frac{a}{b}\right)^{\frac{1}{p}} &\leq \frac{1}{p}\frac{a}{b} + \left(1 - \frac{1}{p}\right) \\ &= \frac{1}{p}\frac{a}{b} + \frac{1}{q}, \\ i.e., a^{\frac{1}{p}} b^{-\frac{1}{p}} &\leq \frac{1}{p}\frac{a}{b} + \frac{1}{q}, \\ i.e., a^{\frac{1}{p}} b^{1-\frac{1}{p}} &\leq \frac{a}{p} + \frac{b}{q}, \\ i.e., a^{\frac{1}{p}} b^{\frac{1}{q}} &\leq \frac{a}{p} + \frac{b}{q}. \end{aligned}$$

Case 2. Let $b \geq a$. Then $\frac{b}{a} \geq 1$. Put $k = \frac{1}{q}$ and we get the result in a similar fashion.

(ii) If $x = 0$, $x_i = 0$, $i = 1, 2, \ldots, n$. Then the result holds trivially. Let $x, y \neq 0$. Put

$$a_i = \left(\frac{|x_i|}{\|x\|_p}\right)^p, b_i = \left(\frac{|y_i|}{\|y\|_q}\right)^q.$$

$$a_i^{\frac{1}{p}} b_i^{\frac{1}{q}} \leq \frac{a_i}{p} + \frac{b_i}{q}, i = 1, 2, \ldots, n,$$

$$i.e., \frac{|x_i||y_i|}{\|x\|_p \|y\|_q} \leq \frac{1}{p}\left(\frac{|x_i|}{\|x\|_p}\right)^p + \frac{1}{q}\left(\frac{|y_i|}{\|y\|_q}\right)^q.$$

Summing the above for $i = 1, 2, \ldots, n$, we get

$$\begin{aligned}
\frac{\sum_{i=1}^{n} |x_i y_i|}{\|x\|_p \|y\|_q} &\leq \frac{1}{p} \frac{1}{\|x\|_p^p} \left(\sum_{i=1}^{n} |x_i|^p\right) \\
&\quad + \frac{1}{q} \frac{1}{\|y\|_q^q} \left(\sum_{i=1}^{n} |y_i|^q\right) \\
&= \frac{1}{p} \frac{1}{\|x\|_p^p} \|x\|_p^p \\
&\quad + \frac{1}{q} \frac{1}{\|y\|_q^q} \|y\|_q^q \\
&= \frac{1}{p} + \frac{1}{q} \\
&= 1,
\end{aligned}$$

$$i.e., \sum_{i=1}^{n} |x_i y_i| \leq \|x\|_p \|y\|_q,$$

establishing Hölder's inequality.

(iii) If $p = 1$, then

$$\begin{aligned}
\|x + y\|_1 &= \sum_{i=1}^{n} |x_i + y_i| \\
&\leq \sum_{i=1}^{n} |x_i| + \sum_{i=1}^{n} |y_i| \\
&= \|x\|_1 + \|y\|_1.
\end{aligned}$$

If $p > 1$,

$$\begin{aligned}
\|x+y\|_p^p &= \sum_{i=1}^{n} |x_i + y_i|^p \\
&= \sum_{i=1}^{n} |x_i + y_i||x_i + y_i|^{p-1} \\
&\leq \sum_{i=1}^{n} (|x_i| + |y_i|)|x_i + y_i|^{p-1} \\
&= \sum_{i=1}^{n} |x_i||x_i + y_i|^{p-1} + \sum_{i=1}^{n} |y_i||x_i + y_i|^{p-1} \\
&= \sum_{i=1}^{n} |x_i||x_i + y_i|^{\frac{p}{q}} + \sum_{i=1}^{n} |y_i||x_i + y_i|^{\frac{p}{q}}, \\
&\qquad \text{using } \frac{1}{p} + \frac{1}{q} = 1 \\
&\leq \|x\|_p \|z\|_q + \|y\|_p \|z\|_q, \\
&\qquad \text{using (ii), where } z = (z_1, z_2, \ldots, z_n), \\
&\qquad z_i = |x_i + y_i|^{\frac{p}{q}}, i = 1, 2, \ldots, n, \\
&= (\|x\|_p + \|y\|_p)\|z\|_q.
\end{aligned}$$

Now,

$$\begin{aligned}
\|z\|_q &= \left( \sum_{i=1}^{n} |z_i|^q \right)^{\frac{1}{q}} \\
&= \left( \sum_{i=1}^{n} |x_i + y_i|^p \right)^{\frac{1}{q}} \\
&= \left( \|x + y|_p^p \right)^{\frac{1}{q}} \\
&= \|x + y|_p^{\frac{p}{q}}.
\end{aligned}$$

Consequently,

$$\|x+y\|_p^p \leq (\|x\|_p + \|y\|_p)\|x+y\|_p^{\frac{p}{q}},$$

$$i.e., \|x+y\|_p^{p-\frac{p}{q}} \leq \|x\|_p + \|y\|_p,$$

$$i.e., \|x+y\|_p \leq \|x\|_p + \|y\|_p, \text{ as } p - \frac{p}{q} = 1.$$

Thus,

$$\|x + y\|_p \leq \|x\|_p + \|y\|_p,$$

for $1 \leq p < \infty$, proving the inequality. □

We now proceed to give some examples of Banach spaces.

**Example 18.** *The set $\mathbb{C}$ of all complex numbers with the usual addition, scalar multiplication, and the norm defined by*

$$\|z\| = |z|, z \in \mathbb{C},$$

*is a Banach space. Similarly, $\mathbb{R}$ is a Banach space too.*

**Example 19.** $\mathbb{C}^n$, $n = 1, 2, \ldots,$ *is a linear space under coordinatewise addition and scalar multiplication. It is a Banach space under the norm*

$$\|x\| = \left\{\sum_{i=1}^{n} |x_i|^2\right\}^{\frac{1}{2}}, x = (x_1, x_2, \ldots, x_n) \in \mathbb{C}^n.$$

*We note that $\|.\|$ is a norm using Lemma 1. It remains to show that $\mathbb{C}^n$ is complete. Note that each element of $\mathbb{C}^n$ can be regarded as a function (or mapping) on the set $\{1, 2, \ldots, n\}$. So, let $\{f_m\}$ be a Cauchy sequence in $\mathbb{C}^n$, so that for any $\epsilon > 0$, we can find a positive integer $n_0$ such that*

$$\|f_m - f_{m'}\| < \epsilon, \text{ for all } m, m' \geq n_0,$$

$$\text{i.e.,} \sum_{i=1}^{n} |f_m(i) - f_{m'}(i)|^2 < \epsilon^2, \text{ for all } m, m' \geq n_0.$$

*Then,*

$$|f_m(i) - f_{m'}(i)|^2 < \epsilon^2,$$

$$\text{i.e.,} |f_m(i) - f_{m'}(i)| < \epsilon, \text{ for all } m, m' \geq n_0$$

$$\text{and for all } i = 1, 2, \ldots, n.$$

*Hence $\{f_m(i)\}$ is a Cauchy sequence in $\mathbb{C}$, for $i = 1, 2, \ldots, n$. Since $\mathbb{C}$ is complete,*

$$f_m(i) \to f(i), m \to \infty, i = 1, 2, \ldots, n.$$

*Let* $f = (f(1), f(2), \ldots, f(n))$*. Then* $f \in \mathbb{C}^n$*. We now claim that*

$$f_m \to f, m \to \infty.$$

*For every* $i = 1, 2, \ldots, n$*, since*

$$f_m(i) \to f(i), m \to \infty,$$

*for* $\epsilon > 0$*, there exists a positive integer* $m_0(i)$*, such that*

$$|f_m(i) - f(i)| < \frac{\epsilon}{\sqrt{n}}, \text{ for all } m \geq m_0(i).$$

*We have*

$$\begin{aligned} \|f_m - f\| &= \left( \sum_{i=1}^{n} |f_m(i) - f(i)|^2 \right)^{\frac{1}{2}} \\ &< \left( \sum_{i=1}^{n} \frac{\epsilon^2}{n} \right)^{\frac{1}{2}} \\ &= \epsilon, \text{ for all } m \geq m_0, \end{aligned}$$

*where* $m_0 = \max_{1 \leq i \leq n} (m_0(i))$*, which proves our claim. Thus* $\mathbb{C}^n$ *is complete and consequently* $\mathbb{C}^n$ *is a Banach space. Similarly* $\mathbb{R}^n$ *is also a Banach space.*

**Example 20.** $\|.\|_p$*, defined in Lemma 2, is a norm and* $\mathbb{C}^n$ *is a normed linear space under this norm. This space is denoted by* $\ell_p^n$*. It is easy to prove that* $\ell_p^n$ *is a Banach space.*

**Example 21.** *Let* $1 \leq p < \infty$*. Let* $\ell_p$ *denote the set of all complex sequences* $x = \{x_n\}$ *such that*

$$\sum_{i=1}^{\infty} |x_i|^p < \infty,$$

*with the norm defined by*

$$\|x\|_p = \left( \sum_{i=1}^{\infty} |x_i|^p \right)^{\frac{1}{p}}.$$

*We can show that* $\ell_p$ *is a Banach space.*

*With respect to the usual operations of coordinatewise addition and scalar multiplication, $\ell_p$ is a linear space. If $x = \{x_n\}$, $y = \{y_n\} \in \ell_p$,*

$$\left(\sum_{i=1}^{n}|x_i + y_i|^p\right)^{\frac{1}{p}} \leq \left(\sum_{i=1}^{n}|x_i|^p\right)^{\frac{1}{p}} + \left(\sum_{i=1}^{n}|y_i|^p\right)^{\frac{1}{p}},$$

*using Lemma 2(iii)*

$$\leq M < \infty, \text{ for every } n = 1, 2, \ldots \tag{1.5}$$

*where*

$$M = \left(\sum_{i=1}^{\infty}|x_i|^p\right)^{\frac{1}{p}} + \left(\sum_{i=1}^{\infty}|y_i|^p\right)^{\frac{1}{p}} > 0.$$

*Thus,*

$$\left(\sum_{i=1}^{\infty}|x_i + y_i|^p\right)^{\frac{1}{p}} < \infty$$

*and so*

$$\sum_{i=1}^{\infty}|x_i + y_i|^p < \infty.$$

*Consequently,*

$$x + y \in \ell_p.$$

*Similarly, $\alpha x \in \ell_p$ and so $\ell_p$ is a linear space. From (1.5), we have*

$$\|x + y\|_p \leq \|x\|_p + \|y\|_p.$$

*The reader can check that $\|x\|_p = 0$ if and only if $x = 0$, $\|\alpha x\|_p = |\alpha|\|x\|_p$. Thus $\ell_p$ is a normed linear space. It remains to show that $\ell_p$ is complete. Now, let $\{x_n\}$ be a Cauchy sequence in $\ell_p$, with*

$$x_n = \{\xi_1^n, \xi_2^n, \ldots, \xi_k^n, \ldots\}, n = 1, 2, \ldots.$$

*For each fixed $k = 1, 2, \ldots$, $\{\xi_k^n\}_{n=1}^{\infty}$ is a Cauchy sequence, because*

$$|\xi_k^n - \xi_k^m| \leq \left(\sum_{k=1}^{\infty}|\xi_k^n - \xi_k^m|^p\right)^{\frac{1}{p}}$$

$$= \|x_n - x_m\|_p$$

$$\to 0, n, m \to \infty.$$

*Since* $\mathbb{C}$ *is complete,* $\{\xi_k^n\}_{n=1}^{\infty}$ *converges to* $\xi_k \in \mathbb{C}$ *(say). Let* $x = \{\xi_1, \xi_2, \ldots, \xi_n, \ldots\}$. *Since* $\{x_n\}$ *is Cauchy, given* $\epsilon > 0$, *there exists a positive integer* $n_0$ *such that*

$$\|x_n - x_m\|_p < \epsilon, \text{ for all } n, m \geq n_0.$$

*Now,*

$$\begin{aligned}\|x_n\|_p &= \|(x_n - x_{n_0}) + x_{n_0}\|_p \\ &\leq \|x_n - x_{n_0}\|_p + \|x_{n_0}\|_p \\ &< \epsilon + \|x_{n_0}\|_p, \text{ for all } n \geq n_0.\end{aligned}$$

*Consequently,*

$$\|x_n\|_p \leq M = \max\{\epsilon + \|x_{n_0}\|_p, \|x_1\|_p, \|x_2\|_p, \ldots, \|x_{n_0-1}\|_p\}, \text{ for all } n = 1, 2, \ldots,$$

$$\text{i.e., } \{\|x_n\|_p\} \text{ is bounded.}$$

*For any* $k, n = 1, 2, \ldots,$

$$\left(\sum_{i=1}^{k} |\xi_i^n|^p\right)^{\frac{1}{p}} \leq \|x_n\|_p \leq M.$$

*Allowing* $n \to \infty$,

$$\left(\sum_{i=1}^{k} |\xi_i|^p\right)^{\frac{1}{p}} \leq M, \text{ for all } k = 1, 2, \ldots,$$

*from which it follows that*

$$\left(\sum_{i=1}^{\infty} |\xi_i|^p\right)^{\frac{1}{p}} \leq M,$$

$$\text{i.e., } x \in \ell_p.$$

*We now claim that*

$$x_n \to x, n \to \infty.$$

*For any* $k = 1, 2, \ldots,$

$$\left(\sum_{i=1}^{k} |\xi_i^n - \xi_i^m|^p\right)^{\frac{1}{p}} \leq \|x_n - x_m\|_p$$

$$< \epsilon, \text{ for all } n, m \geq n_0.$$

*Keeping* $k$ *fixed and allowing* $m \to \infty$, *we have*

$$\left(\sum_{i=1}^{k} |\xi_i^n - \xi_i|^p\right)^{\frac{1}{p}} \leq \epsilon, \text{ for all } k = 1, 2, \ldots \text{ and } n \geq n_0.$$

*Consequently,*

$$\left(\sum_{i=1}^{\infty} |\xi_i^n - \xi_i|^p\right)^{\frac{1}{p}} \leq \epsilon, \text{ for all } n \geq n_0,$$

$$\text{i.e.,} \|x_n - x\|_p \leq \epsilon, \text{ for all } n \geq n_0,$$

$$\text{i.e.,} x_n \to x, n \to \infty,$$

*completing the proof.*

**Example 22.** *Let* $\ell_\infty$ *be the set of all bounded complex sequences. For* $x = \{x_n\} \in \ell_\infty$, *define*

$$\|x\|_\infty = \sup_{i \geq 1} |x_i|. \tag{1.6}$$

*The reader can check that* $\ell_\infty$ *is a Banach space. Similarly, it could be proved that* $c_0$ *and* $c$, *the sets of all null and convergent complex sequences, respectively, are Banach spaces under the norm (1.6).*

**Example 23.** *Let* $C(X)$ *denote the set of all bounded, complex-valued, continuous functions defined on the topological space* $X$. $C(X)$ *is a linear space under usual operations of addition and scalar multiplication. For* $f \in C(X)$, *define*

$$\|f\| = \sup_{x \in X} |f(x)|.$$

*One can verify that* $\|.\|$ *is a norm on* $C(X)$. *It remains to prove that* $C(X)$ *is complete. Toward this end, let* $\{f_n\}$ *be a Cauchy sequence in* $C(X)$. *So, given*

*$\epsilon > 0$, we can find a positive integer $N$ such that*

$$\|f_n - f_m\| < \epsilon, \text{ for all } n, m \geq N,$$

$$\text{i.e., } \sup_{x \in X} |f_n(x) - f_m(x)| < \epsilon, \text{ for all } n, m \geq N,$$

$$\text{i.e., } |f_n(x) - f_m(x)| < \epsilon, \text{ for all } n, m \geq N \text{ and for all } x \in X.$$

*Hence the sequence $\{f_n\}$ is a uniform Cauchy sequence of continuous functions and so converges uniformly to some function $f$. Note that $f$ is continuous so that $f \in C(X)$. Again, by uniform convergence,*

$$\|f_n - f\| \to 0, n \to \infty,$$

$$\text{i.e., } f_n \to f, n \to \infty,$$

*where $f \in C(X)$, completing the proof.*

Note that if $X$ is compact, $C(X)$ is the space of all complex-valued continuous functions on $X$.

**Definition 19.** *Let $X$ be a normed linear space. We say that the infinite series $\sum_{n=1}^{\infty} x_n$, $x_n \in X$, $n = 1, 2, \ldots$ "converges" to $s \in X$ if*

$$s_n \to s, n \to \infty,$$

*where $s_n = \sum_{i=1}^{n} x_i$, $n - 1, 2, \ldots$. In this case, we write*

$$\sum_{n=1}^{\infty} x_n = s.$$

*A series $\sum_{n=1}^{\infty} x_n$ is said to "converge absolutely" if $\sum_{n=1}^{\infty} \|x_n\| < \infty$.*

*In $\mathbb{R}$ or $\mathbb{C}$, every absolutely convergent series converges.*

## Exercises

1. Let $x = (1, 2) \in \mathbb{R}^2$. Find $\|x\|_p$, for $p = 1, 2, 4$.

2. If $X, Y$ are normed linear spaces, prove that $X \times Y$ is a linear space under addition and scalar multiplication defined coordinatewise. For $(x, y) \in X \times Y$, define

$$\|(x, y)\| = \|x\| + \|y\|.$$

Prove that $\|.\|$ is a norm on $X \times Y$. Is $X \times Y$ complete under the above norm? Justify your answer. Prove also that $X \times Y$ is a Banach space if $X$ and $Y$ are Banach spaces.

3. Let $0 < p < 1$ and

$$\|x\|_p = \left(\sum_{i=1}^{n} |x_i|^p\right)^{\frac{1}{p}}, x = (x_1, x_2, \ldots, x_n) \in \mathbb{C}^n.$$

Prove that $\|.\|_p$ is not a norm on $\mathbb{C}^n$, if $n \geq 2$.

4. In a Banach space, prove that every absolutely convergent series is convergent. Is the converse true?
Justify your answer.

5. Prove that a normed linear space $X$ is complete if and only if every absolutely convergent series is convergent.

6. Let $X \neq \{0\}$ be a normed linear space. Prove that $X$ is a Banach space if and only if $\{x \in X/\|x\| = 1\}$ is complete.

7. In a normed linear space $X$, let $\sum_{n=1}^{\infty} |x_n - x_{n+1}| < \infty$, $x_n \in X$, $n = 1, 2, \ldots$. Is $\{x_n\}$ Cauchy? Is $\{x_n\}$ convergent? Justify your answers.

8. Let $S$ be a dense subset of a normed linear space $X$. Prove that every element of $X$ can be written as the sum of an absolutely convergent series whose terms are in $S$.

9. In a normed linear space, prove that a Cauchy sequence that has a convergent subsequence converges to the same limit.

10. Prove that a closed subspace $Y$ of a Banach space $X$ is itself a Banach space.

# Chapter 2

# Linear Transformations, Linear Functionals, Convexity

## 2.1 Factor spaces, linear transformation, and continuity of a linear transformation

**Theorem 6.** *The closure of a linear subspace of a normed linear space $X$ is itself a subspace.*

*Proof.* Let $M$ be a linear subspace of the normed linear space $X$. We will prove that $\alpha x + \beta y \in \overline{M}$, the closure of $M$, whenever $x, y \in \overline{M}$, $\alpha, \beta \in \mathbb{C}$. Leaving out the trivial part, let $\alpha, \beta \neq 0$. Since $x, y \in \overline{M}$, given $\epsilon > 0$, there exist $x_1, y_1 \in M$ such that

$$\|x - x_1\| < \frac{\epsilon}{2|\alpha|}, \|y - y_1\| < \frac{\epsilon}{2|\beta|}.$$

Since $x_1, y_1 \in M$ and $M$ is a linear subspace $X$, $\alpha x_1 + \beta y_1 \in M$. Thus,

$$\begin{aligned}
\|(\alpha x + \beta y) - (\alpha x_1 + \beta y_1)\| &= \|\alpha(x - x_1) + \beta(y - y_1)\| \\
&\leq \|\alpha(x - x_1)\| + \|\beta(y - y_1)\| \\
&= |\alpha|\|x - x_1\| + |\beta|\|y - y_1\| \\
&< |\alpha|\frac{\epsilon}{2|\alpha|} + |\beta|\frac{\epsilon}{2|\beta|} \\
&= \frac{\epsilon}{2} + \frac{\epsilon}{2} \\
&= \epsilon,
\end{aligned}$$

proving that $\alpha x + \beta y \in \overline{M}$. The cases $\alpha = 0$, $\beta \neq 0$ and $\alpha \neq 0$, $\beta = 0$ can be similarly discussed, completing the proof. □

**Theorem 7.** *Let $M$ be a closed subspace of a normed linear space $X$. For $x + M \in X/M$, define*

$$\|x + M\| = \inf_{m \in M} \|x + m\|. \tag{2.1}$$

*Then (2.1) defines a norm on $X/M$. Further, if $X$ is a Banach space, then $X/M$ is a Banach space too.*

*Proof.* If $x + M = 0$, $x + M = M$, since $M$ is the zero of $X/M$. Hence,

$$0 \leq \|x + M\| = \|M\| = \inf_{m \in M} \|m\| \leq \|0\| = 0,$$

which implies that

$$\|x + M\| = 0.$$

Conversely, let $\|x+M\| = 0$, i.e., $\inf_{m \in M} \|x+m\| = 0$. So there exists a sequence $\{m_n\}$ in $M$ such that

$$\|x + m_n\| \to 0, n \to \infty.$$

Hence, given $\epsilon > 0$, there exists a positive integer $n_0$ such that

$$\|x + m_n\| < \epsilon, \text{ for all } n \geq n_0.$$

Let $U_\epsilon(x)$ denote the open ball with $x$ as the center and $\epsilon$ as the radius. Then

$$m_n \in U_\epsilon(x) \cap M, \text{ for all } n \geq n_0.$$

Therefore, $U_\epsilon(x) \cap M \neq \phi$. Hence $x \in \overline{M} = M$, since $M$ is closed. Thus $x + M = M$, i.e., $x + M = 0$, $M$ being the zero of $X/M$.

If $\alpha = 0$, $\|\alpha(x + M)\| = |\alpha|\|x + M\|$ is easily verified. If $\alpha \neq 0$,

$$\begin{aligned}
\|\alpha(x + M)\| = \|\alpha x + M\| &= \inf_{m \in M} \|\alpha x + m\| \\
&= \inf_{m \in M} \|\alpha(x + \frac{m}{\alpha})\| \\
&= |\alpha| \inf_{m \in M} \|x + \frac{m}{\alpha}\| \\
&= |\alpha|\|x + M\|.
\end{aligned}$$

There exist sequences $\{m_n\}$, $\{m'_n\}$ in $M$ such that

$$\|x + M\| = \inf_{m \in M} \|x + m\| = \lim_{n \to \infty} \|x + m_n\|$$

and

$$\|y + M\| = \lim_{n \to \infty} \|y + m'_n\|.$$

Now,

$$\begin{aligned} \|x + y + M\| &\leq \|x + y + (m_n + m'_n)\| \\ &= \|(x + m_n) + (y + m'_n)\| \\ &\leq \|x + m_n\| + \|y + m'_n\|. \end{aligned}$$

Taking the limit as $n \to \infty$, we get

$$\|x + y + M\| \leq \|x + M\| + \|y + M\|.$$

Thus (2.1) defines a norm on $X/M$.

Now, let $X$ be complete. To prove that $X/M$ is also complete, let $\{x_n + M\}$ be a Cauchy sequence in $X/M$. We can choose positive integers $n_1, n_2, n_3, \ldots$ such that

$$\begin{aligned} \|(x_{n_2} + M) - (x_{n_1} + M)\| &< \frac{1}{2}, \\ \|(x_{n_3} + M) - (x_{n_2} + M)\| &< \frac{1}{2^2}, \end{aligned}$$

and so on. Let $y_1$ be a fixed element of $x_{n_1} + M$. Then, there exists an element $y_2 \in x_{n_2} + M$ such that

$$\|y_2 - y_1\| < \frac{1}{2}, \text{ using the definition of infimum.}$$

Similarly, we can choose $y_3 \in x_{n_3} + M$ such that

$$\|y_3 - y_2\| < \frac{1}{2^2}$$

and so on. Now, for $n > m$,

$$\begin{aligned}\|y_n - y_m\| &= \|(y_n - y_{n-1}) + (y_{n-1} - y_{n-2}) + \cdots + (y_{m+1} - y_m)\| \\ &\leq \|y_n - y_{n-1}\| + \|y_{n-1} - y_{n-2}\| + \cdots + \|y_{m+1} - y_m\| \\ &< \frac{1}{2^m} + \frac{1}{2^{m+1}} + \cdots + \frac{1}{2^{n-1}} \\ &\to 0, m \to \infty, \text{ since } \sum_{i=1}^{\infty} \frac{1}{2^i} \text{ converges,}\end{aligned}$$

so that $\{y_n\}$ is a Cauchy sequence in $X$. Since $X$ is complete, there exists $y \in X$ such that

$$y_n \to y, n \to \infty.$$

We now claim that

$$x_{n_k} + M \to y + M, k \to \infty, \tag{2.2}$$

because

$$\begin{aligned}\|(x_{n_k} + M) - (y + M)\| &\leq \|y_k - y\| \\ &\to 0, k \to \infty.\end{aligned}$$

Using (2.2) and the fact that $\{x_n + M\}$ is a Cauchy sequence, we have

$$\begin{aligned}\|(x_n + M) - (y + M)\| &\leq \|(x_n + M) - (x_{n_k} + M)\| \\ &\quad + \|(x_{n_k} + M) - (y + M)\| \\ &\to 0, n, k \to \infty,\end{aligned}$$

proving that

$$x_n + M \to y + M, n \to \infty,$$

completing the proof. □

$X/M$ is called the "factor space $X$ modulo $M$."

**Definition 20.** *Let $X, Y$ be normed linear spaces over $\mathbb{C}$. A mapping $T : X \to Y$ is called a "linear transformation" if*

$$T(\alpha x + \beta y) = \alpha T(x) + \beta T(y),$$

*for all* $x, y \in X$ *and* $\alpha, \beta \in \mathbb{C}$. *We can check that*

$$T(X) = \{T(x)/x \in X\}$$

*is a subspace of* $Y$.

**Definition 21.** *A mapping* $T : X \to Y$ *is said to be "continuous" if whenever* $x_n \to x$, $n \to \infty$ *in* $X$,

$$T(x_n) \to T(x), n \to \infty \text{ in } Y.$$

**Theorem 8.** *Let* $X, Y$ *be normed linear spaces and* $T$ *be a linear transformation from* $X$ *to* $Y$. *Then the following statements are equivalent:*

*(i)* $T$ *is continuous;*

*(ii)* $T$ *is continuous at 0;*

*(iii)* $T$ *is bounded, i.e.,* $\|T(x)\| \leq M\|x\|$, $M > 0$, *for all* $x \in X$;

*and*

*(iv) if* $S = \{x \in X/\|x\| \leq 1\}$, *then* $T(S)$ *is bounded,*

*i.e.,* $\|T(x)\| \leq M$, $M > 0$, *for all* $x \in S$.

*Proof.* It is clear that (i) implies (ii). Conversely, let $x_n \to x$, $n \to \infty$ in $X$, i.e., $x_n - x \to 0$, $n \to \infty$. Since $T$ is continuous at 0,

$$T(x_n - x) \to T(0) = 0, n \to \infty,$$

$$\text{i.e.,} T(x_n) - T(x) \to 0, n \to \infty,$$

$$\text{i.e.,} T(x_n) \to T(x), n \to \infty,$$

*i.e.,* $T$ is continuous,

so that (ii) implies (i). Next, we prove that (ii) implies (iii). Suppose (ii) holds and (iii) does not hold. Since $T$ is not bounded, given $\epsilon > 0$, there exists

$x_n \in X$, $x_n \neq 0$, $n = 1, 2, \ldots$ such that

$$\|T(x_n)\| > n\|x_n\|,$$
$$i.e., \frac{\|T(x_n)\|}{n\|x_n\|} > 1,$$
$$i.e., \left\|T\left(\frac{x_n}{n\|x_n\|}\right)\right\| > 1.$$

Let $y_n = \frac{x_n}{n\|x_n\|}$. Then

$$\begin{aligned}\|y_n - 0\| = \|y_n\| &= \left\|\frac{x_n}{n\|x_n\|}\right\| \\ &= \frac{1}{n} \to 0, n \to \infty,\end{aligned}$$

so that $y_n \to 0$, $n \to \infty$. However, $\|T(y_n)\| > 1$ so that $T(y_n) \not\to 0$, $n \to \infty$, which contradicts (ii). Thus (ii) implies (iii). Conversely, let $T$ be bounded, i.e., $\|T(x)\| \leq M\|x\|$, $M > 0$, for all $x \in X$. Let $x_n \to 0$, $n \to \infty$. Now,

$$\begin{aligned}\|T(x_n) - T(0)\| &= \|T(x_n - 0)\| \\ &= \|T(x_n)\| \\ &\leq M\|x_n\| \\ &\to 0, n \to \infty,\end{aligned}$$

so that $T$ is continuous at 0. Consequently, (iii) implies (ii). (iii) implies (iv), because, if $x \in S$,

$$\|T(x)\| \leq M\|x\| \leq M, \text{ since } \|x\| \leq 1,$$

so that $T(S)$ is bounded. Finally, we prove that (iv) implies (iii). Let $T(S)$ be bounded. If $x = 0$, $T(x) = 0$ so that

$$\|T(x)\| \leq M\|x\|$$

is trivially true. Let $x \in X$, $x \neq 0$. Then $\frac{x}{\|x\|} \in S$, since $\left\|\frac{x}{\|x\|}\right\| = 1$. By hypothesis, $\left\|T\left(\frac{x}{\|x\|}\right)\right\| \leq M$, i.e., $\frac{1}{\|x\|}\|T(x)\| \leq M$, i.e., $\|T(x)\| \leq M\|x\|$, i.e., (iii) holds, completing the proof of the theorem. □

Note that, in view of Theorem 8, a linear transformation $T : X \to Y$ is continuous if and only if it is bounded.

**Theorem 9.** *Let $X, Y$ be normed linear spaces over $\mathbb{C}$, $X \neq \{0\}$. Let $T : X \to Y$ be a bounded linear transformation of $X$ to $Y$. Then*

$$\sup_{\|x\|=1} \|T(x)\| = \sup_{\|x\|\leq 1} \|T(x)\| = \sup_{x\neq 0} \frac{\|T(x)\|}{\|x\|} < \infty.$$

*Proof.* Let

$$M_1 = \sup_{\|x\|=1} \|T(x)\|;$$

$$M_2 = \sup_{\|x\|\leq 1} \|T(x)\|;$$

*and*

$$M_3 = \sup_{x\neq 0} \frac{\|T(x)\|}{\|x\|}.$$

Since $T$ is bounded,

$$\|T(x)\| \leq M\|x\|, M > 0, \text{ for all } x \in X.$$

If $\|x\| = 1$, $\|T(x)\| \leq M$ and so $M_1 < \infty$.

If $\|x\| \leq 1$, $\|T(x)\| \leq M$ and so $M_2 < \infty$.

If $x \neq 0$, $\frac{\|T(x)\|}{\|x\|} \leq M$ and so $M_3 < \infty$.

Now, it is clear that $M_1 \leq M_2$. Let $\|x\| \leq 1$. If $x = 0$, $\|T(x)\| = 0 \leq M$; if $x \neq 0$, $\left\|\frac{x}{\|x\|}\right\| = 1$ and so

$$\left\|T\left(\frac{x}{\|x\|}\right)\right\| \leq M_1,$$

$$i.e., \frac{1}{\|x\|}\|T(x)\| \leq M_1,$$

$$i.e., \|T(x)\| \leq M_1\|x\|,$$

$$i.e., \|T(x)\| \leq M_1, \text{ for all } x \text{ such that } \|x\| \leq 1$$

and so

$$M_2 = \sup_{\|x\|\leq 1} \|T(x)\| \leq M_1.$$

Consequently, $M_1 = M_2$.

For $x$ with $\|x\| = 1$,

$$\|T(x)\| = \frac{\|T(x)\|}{\|x\|} \leq M_3$$

and so

$$\sup_{\|x\|=1} \|T(x)\| \leq M_3,$$

$$i.e., M_1 \leq M_3.$$

Let $x \neq 0$. Then

$$\frac{\|T(x)\|}{\|x\|} = \left\| T\left(\frac{x}{\|x\|}\right) \right\| \leq M_1,$$

since $\left\| \frac{x}{\|x\|} \right\| = 1$. Thus,

$$M_3 = \sup_{x \neq 0} \frac{\|T(x)\|}{\|x\|} \leq M_1$$

and consequently $M_1 = M_3$. We have thus proved that $M_1 = M_2 = M_3$. □

This number is called the "norm of $T$" and denoted, as usual, by $\|T\|$. Incidentally, we note that

$$\|T\| = \inf\{M/\|T(x)\| \leq M\|x\|, \text{ for all } x \in X\}.$$

We also note that

$$\|T(x)\| \leq \|T\|\|x\|, \text{ for all } x \in X,$$

because, if $x = 0$, the result is trivially valid; if $x \neq 0$,

$$\|T\| = \sup_{x \neq 0} \frac{\|T(x)\|}{\|x\|}.$$

Thus,

$$\frac{\|T(x)\|}{\|x\|} \leq \|T\|, \text{ for all } x \neq 0,$$

$$i.e., \|T(x)\| \leq \|T\|\|x\|, \text{ for all } x \neq 0.$$

Consequently,

$$\|T(x)\| \leq \|T\|\|x\|, \text{ for all } x \in X.$$

Let $B(X, Y)$ denote the set of all bounded linear transformations of $X$ to $Y$, where $X, Y$ are normed linear spaces over $\mathbb{C}$. Define, as usual,

$$(T_1 + T_2)(x) = T_1(x) + T_2(x);$$
$$(\alpha T)(x) = \alpha T(x),$$

for $T, T_1, T_2 \in B(X, Y)$ and $\alpha \in \mathbb{C}$. It is clear that $T_1 + T_2$, $\alpha T$ are linear transformations. Since $T_1, T_2$ are bounded, there exists $M > 0$ such that

$$\|T_1(x)\| \leq M\|x\|$$

and

$$\|T_2(x)\| \leq M\|x\|, \text{ for all } x \in X.$$

Therefore,

$$\begin{aligned}\|(T_1 + T_2)(x)\| &= \|T_1(x) + T_2(x)\| \\ &\leq \|T_1(x)\| + \|T_2(x)\| \\ &\leq 2M\|x\|, \text{ for all } x \in X,\end{aligned}$$

so that $T_1 + T_2$ is bounded. Similarly, $\alpha T$ is bounded, proving that $T_1 + T_2, \alpha T \in B(X, Y)$. Thus $B(X, Y)$ is a linear space. Now,

$$\|T\| = \sup_{x \neq 0} \frac{\|T(x)\|}{\|x\|}.$$

We will prove that $\|.\|$ is a norm on $B(X, Y)$. It is clear that if $T = 0$, then $\|T\| = 0$. Conversely, let

$$\begin{aligned}&\|T\| = 0, \\ &i.e., \sup_{x \neq 0} \frac{\|T(x)\|}{\|x\|} = 0, \\ &i.e., \frac{\|T(x)\|}{\|x\|} = 0, \text{ for all } x \in X, \\ &i.e., \|T(x)\| = 0, \text{ for all } x \in X, \\ &i.e., T(x) = 0, \text{ for all } x \in X, \\ &i.e., T = 0.\end{aligned}$$

The reader can easily check that $\|\alpha T\| = |\alpha|\|T\|$. Now,

$$\begin{aligned}\|T_1 + T_2\| &= \sup_{\|x\|=1} \|(T_1 + T_2)(x)\| \\ &\leq \sup_{\|x\|=1} (\|T_1(x)\| + \|T_2(x)\|) \\ &= \sup_{\|x\|=1} \|T_1(x)\| + \sup_{\|x\|=1} \|T_2(x)\| \\ &= \|T_1\| + \|T_2\|,\end{aligned}$$

proving that $B(X, Y)$ is a normed linear space.

**Theorem 10.** *If $Y$ is a Banach space, then $B(X, Y)$ is also a Banach space.*

*Proof.* Let $\{T_n\}$ be a Cauchy sequence in $B(X, Y)$, where $Y$ is complete. Let $x \in X$. Then

$$\begin{aligned}\|T_n(x) - T_m(x)\| &= \|(T_n - T_m)(x)\| \\ &\leq \|T_n - T_m\|\|x\| \\ &\to 0, n, m \to \infty,\end{aligned}$$

so that, for every $x \in X$, $\{T_n(x)\}$ is a Cauchy sequence in $Y$, which is complete. Thus,

$$T_n(x) \to T(x) \in Y, n \to \infty,$$

so that with every $x \in X$, we associate a unique $T(x) \in Y$. In other words, $T$ is a mapping of $X$ to $Y$. $T$ is linear, because

$$\begin{aligned}T(\alpha x + \beta y) &= \lim_{n\to\infty} T_n(\alpha x + \beta y) \\ &= \lim_{n\to\infty} [\alpha T_n(x) + \beta T_n(y)] \\ &= \alpha \lim_{n\to\infty} T_n(x) + \beta \lim_{n\to\infty} T_n(y) \\ &= \alpha T(x) + \beta T(y).\end{aligned}$$

We next claim that $T$ is bounded. Since $\{T_n\}$ is a Cauchy sequence, it is bounded, i.e., for some $M > 0$,

$$\|T_n\| \leq M, n = 1, 2, \ldots.$$

Now,

$$\|T(x)\| = \|\lim_{n\to\infty} T_n(x)\|$$
$$= \lim_{n\to\infty} \|T_n(x)\|, \text{ since } \|.\| \text{ is continuous.}$$

$$\|T_n(x)\| \leq \|T_n\|\|x\|$$
$$\leq M\|x\|, \text{ for all } n = 1, 2, \ldots.$$

Therefore,

$$\lim_{n\to\infty} \|T_n(x)\| \leq M\|x\|$$

and thus,

$$\|T(x)\| \leq M\|x\|, \text{ for all } x \in X.$$

Consequently, $T$ is bounded and so $T \in B(X, Y)$. We further claim that

$$T_n \to T, n \to \infty.$$

Let $\|x\| \leq 1$. Then,

$$\|T_n(x) - T_m(x)\| = \|(T_n - T_m)(x)\|$$
$$\leq \|T_n - T_m\|\|x\|$$
$$\leq \|T_n - T_m\|, \text{ since } \|(x)\| \leq 1$$
$$< \epsilon, \text{ for all } n, m \geq n_0,$$

$n_0$ being a positive integer, since $\{T_n\}$ is Cauchy. Let $n \geq n_0$. Allowing $m \to \infty$, we have

$$\|T_n(x) - T(x)\| \leq \epsilon, \text{ for all } n \geq n_0,$$

so that

$$\|T_n - T\| = \sup_{\|x\|\leq 1} \|T_n(x) - T(x)\|$$
$$\leq \epsilon, \text{ for all } n \geq n_0,$$
$$i.e., T_n \to T, n \to \infty \text{ in } B(X, Y).$$

In other words, $B(X, Y)$ is complete, proving that $B(X, Y)$ is a Banach space. The proof of the theorem is now complete. □

## 2.2 Linear functionals, bounded linear functionals, conjugate space

In Section 1.1, we defined a "linear functional." We shall now record a few results about linear functionals and then about bounded linear functionals.

**Definition 22.** *Let $f$ be a linear functional on $X$. Let $\aleph(f)$ denote the set of all elements $x \in X$ such that $f(x) = 0$. $\aleph(f)$ is called the "null space" of $f$.*

It is clear that $\aleph(f)$ is a subspace of $X$, because if $x, y \in \aleph(f)$ and $\alpha, \beta \in \mathbb{C}$ (or $\mathbb{R}$),

$$\begin{aligned} f(\alpha x + \beta y) &= \alpha f(x) + \beta f(y) \\ &= 0 + 0 \\ &= 0, \end{aligned}$$

so that $\alpha x + \beta y \in \aleph(f)$.

**Definition 23.** *A linear functional $f$ on $X$ is said to be "non-trivial" if there exists $x \in X$ such that $f(x) \neq 0$.*

The following result gives a kind of representation of the elements of $X$.

**Theorem 11.** *Let $x_0 \in X - \aleph(f)$ be a fixed element. Then any $x \in X$ can be uniquely represented in the form*

$$x = \alpha x_0 + y, \ \textit{where } y \in \aleph(f).$$

*Proof.* Since $x_0 \in X - \aleph(f)$, $f(x_0) \neq 0$ and so $x_0 \neq 0$. Let us suppose that $f(x_0) = 1$, since, otherwise, we have to consider $x_0' = \frac{x_0}{f(x_0)}$, noting that $f(x_0') = \frac{f(x_0)}{f(x_0)} = 1$. Now, for any given $x \in X$, let

$$y = x - \alpha x_0, \text{ where } \alpha = f(x).$$

Then, $y \in \aleph(f)$, since

$$\begin{aligned} f(y) &= f(x - \alpha x_0) \\ &= f(x) - \alpha f(x_0) \\ &= f(x) - \alpha, \text{ since } f(x_0) = 1 \\ &= 0. \end{aligned}$$

Thus,

$$x = y + \alpha x_0, \text{ where } y \in \aleph(f).$$

Note that the above representation of $x$ is unique, since, if

$$x = y' + \alpha' x_0, \text{ where } y' \in \aleph(f),$$

then

$$0 = y - y' + (\alpha - \alpha')x_0,$$

$$i.e.,(\alpha - \alpha')x_0 = y' - y.$$

If $\alpha = \alpha'$, then $y' = y$. If $\alpha \neq \alpha'$, then

$$x_0 = \frac{y' - y}{\alpha - \alpha'} \in \aleph(f),$$

contradicting the choice of $x_0$, completing the proof. □

We have already introduced a linear transformation from a linear space $X$ to another. Since a linear functional can be looked at as a linear transformation from a linear space $X$ to its underlying field, all the properties studied in the context of linear transformations continue to hold for linear functionals too. We now list the analogous results, noting that whenever a norm appeared in the discussion of bounded linear transformations, it should be replaced by the absolute value in the discussion of bounded linear functionals.

**Definition 24.** *A linear functional $f$, defined on a normed linear space $X$, is said to be "bounded" if there exists $M > 0$ such that*

$$|f(x)| \leq M\|x\|, \textit{ for all } x \in X.$$

**Definition 25.** *A linear functional $f$, defined on a normed linear space $X$, is said to be "continuous" at $x_0 \in X$, if*

$$f(x_n) \to f(x_0), n \to \infty,$$

*whenever*

$$x_n \to x_0, n \to \infty \text{ in } X.$$

**Theorem 12.** *Let $f$ be a linear functional, defined on a normed linear space $X$. Then $f$ is continuous at $x_0 \in X$ if and only if $f$ is continuous at every point of $X$.*

**Theorem 13.** *A linear functional, defined on a normed linear space $X$, is bounded if and only if it is continuous.*

Let $f, g$ be bounded linear functionals defined on the normed linear space $X$. We define

$$f = g$$

if

$$f(x) = g(x), \text{ for all } x \in X;$$

$$(f+g)(x) = f(x) + g(x), \text{ for all } x \in X;$$

*and*

$$(\alpha f)(x) = \alpha f(x), \text{ for all } x \in X, \alpha \in \mathbb{C} \text{ or } \mathbb{R}.$$

Under the above operations, the set of all bounded linear functionals defined on $X$ is a linear space. Define the norm of $f$ by

$$\|f\| = \sup_{x \neq 0} \frac{|f(x)|}{\|x\|}, x \in X.$$

The set of all bounded linear functionals defined on $X$ is now a normed linear space under the above norm. We also note that there are other equivalent ways of defining the norm on this linear space.

(i) $\|f\| = \inf_M \{M / |f(x)| \leq M\|x\|, \text{ for all } x \in X\};$

(ii) $\|f\| = \sup_{\|x\| \leq 1} |f(x)|$

and

(iii) $\|f\| = \sup_{\|x\|=1} |f(x)|$.

**Definition 26.** *The normed linear space, consisting of all bounded linear functionals defined on a normed linear space $X$, is called the "conjugate or continuous dual space" of $X$ and is denoted by $\tilde{X}$ or $X^*$.*

Since $\mathbb{R}$ and $\mathbb{C}$ are complete, we have the following result.

**Theorem 14.** *Given a normed linear space $X$, its conjugate space $\tilde{X}$ is always a Banach space.*

The above result is true regardless of whether $X$ is a Banach space or not.

---

## 2.3 Some results on finite-dimensional spaces

**Definition 27.** *Let $\|.\|_1, \|.\|_2$ be two norms defined on a linear space $X$. $\|.\|_1$ is said to be equivalent to $\|.\|_2$, written as $\|.\|_1 \sim \|.\|_2$, if there exist positive numbers $a$ and $b$ such that*

$$a\|x\|_1 \leq \|x\|_2 \leq b\|x\|_1, \text{ for all } x \in X.$$

*Note that $\sim$ is an equivalence relation.*

The following result shows that in the case of finite-dimensional linear space $X$, the conjugate space of $X$, is the same as the space of all linear functionals defined on $X$.

**Theorem 15.** *If $X$ is a finite-dimensional linear space, then all linear functionals defined on $X$ are bounded.*

*Proof.* Let $f$ be a linear functional defined on $X$. Let $\{x_1, x_2, \ldots, x_n\}$ be a basis of $X$ and $x \in X$. We can now find unique scalars $\alpha_1, \alpha_2, \ldots, \alpha_n$, associated with $x$, such that

$$x = \sum_{i=1}^{n} \alpha_i x_i.$$

We now have

$$\begin{aligned} |f(x)| &= \left| f\left(\sum_{i=1}^{n} \alpha_i x_i\right)\right| \\ &= \left|\sum_{i=1}^{n} \alpha_i f(x_i)\right| \\ &\le \sum_{i=1}^{n} |\alpha_i||f(x_i)| \\ &\le \|x\|_0 \sum_{i=1}^{n} |f(x_i)|, \end{aligned}$$

where

$$\|x\|_0 = \max_{1 \le i \le n} |\alpha_i|.$$

The reader can verify that $\|.\|_0$ is a norm on $X$. It is called the "zeroth norm" on $X$. If

$$M = \sum_{i=1}^{n} |f(x_i)|,$$

then

$$|f(x)| \le M\|x\|_0,$$

so that $f$ is bounded under the norm $\|.\|_0$. To complete the proof, it suffices to show that on a finite-dimensional space $X$, all norms are equivalent.

Let $\|.\|$ be any norm on $X$. We will show that $\|.\|$ is equivalent to the zeroth norm $\|.\|_0$. Now,

$$\begin{aligned}\|x\| &= \left\|\sum_{i=1}^{n} \alpha_i x_i\right\| \\ &\leq \sum_{i=1}^{n} |\alpha_i| \|x_i\| \\ &\leq \|x\|_0 \sum_{i=1}^{n} \|x_i\| \\ &= b\|x\|_0,\end{aligned}$$

where $b = \sum_{i=1}^{n} \|x_i\|$, so that

$$\|x\| \leq b\|x\|_0.$$

We shall now prove that there exists a real number $a > 0$ such that

$$a\|x\|_0 \leq \|x\|. \tag{2.3}$$

We prove the result by induction on the dimension of $X$. Note that (2.3) holds trivially for a 1-dimensional space $X$ with basis $x_1$ (say). Any $x \in X$ can be written as

$$x = \alpha_1 x_1, \text{ for a unique scalar } \alpha_1,$$

so that

$$\begin{aligned}\|x\| &= \|\alpha_1 x_1\| \\ &= |\alpha_1| \|x_1\| \\ &= \|x\|_0 \|x_1\| \\ &= a\|x\|_0, \text{ where } a = \|x_1\|.\end{aligned}$$

Hence (2.3) holds in this case. Let us assume that the result holds for all spaces $X$ of dimension $\leq (n-1)$. Now, let $dim\ X = n$ with $\{x_1, x_2, \ldots, x_n\}$ as a basis. Let

$$M = [\{x_1, x_2, \ldots, x_{n-1}\}]$$

be the subspace of $X$ spanned by the first $(n-1)$ basis vectors. Then $\|.\| \sim \|.\|_0$ on $M$, by assumption. Let $\{y_n\}$ be a Cauchy sequence from $M$ under $\|.\|$. Note that $\{y_n\}$ is also a Cauchy sequence under $\|.\|_0$. Now,

$$y_i = \alpha_1^{(i)} x_1 + \alpha_2^{(i)} x_2 + \cdots + \alpha_{n-1}^{(i)} x_{n-1}, i = 1, 2, \ldots.$$

$$\|y_n - y_m\|_0 = \max_{1 \le j \le (n-1)} |\alpha_j^{(n)} - \alpha_j^{(m)}|.$$

Since $\|y_n - y_m\|_0 \to 0,\ m, n \to \infty$,

$$|\alpha_j^{(n)} - \alpha_j^{(m)}| \to 0, m, n \to \infty, \text{ for } j = 1, 2, \ldots, (n-1).$$

Since $\mathbb{R}$ or $\mathbb{C}$ is complete,

$$\alpha_j^{(m)} \to \alpha_j, m \to \infty, \text{ for all } j = 1, 2, \ldots, (n-1).$$

Let

$$y = \sum_{j=1}^{n-1} \alpha_j x_j.$$

Note that

$$y_m \to y, m \to \infty \text{ with respect to } \|.\|_0.$$

This implies that

$$y_m \to y, m \to \infty \text{ under the norm } \|\|.$$

In other words, $M$ is complete with respect to any $\|.\|$.

It now follows that $M$ is closed. Consider

$$x_n + M = \{x_n + z / z \in M\}.$$

Consider the mapping

$$M \to x_n + M$$

$$z \to x_n + z, z \in M.$$

This mapping is an isometry. Since $M$ is closed, $x_n + M$ is also closed, so that $(x_n + M)'$, i.e., the complement of $x_n + M$ is open. We now claim that $0 \notin x_n + M$, because if $0 \in x_n + M$, we can write

$$0 = x_n + \beta_1 x_1 + \beta_2 x_2 + \cdots + \beta_{n-1} x_{n-1},$$

for some scalars $\beta_i$, $i = 1, 2, \ldots, (n-1)$, which is a contradiction. Thus, $0 \in (x_n + M)'$, which is open. So there exists some neighborhood of 0, lying entirely inside $(x_n + M)'$, i.e., there exists $c_n > 0$ such that $\|x\| \geq c_n$, for all $x \in x_n + M$. Thus, for some scalars $\alpha_i$, $i = 1, 2, \ldots, (n-1)$,

$$x = \alpha_1 x_1 + \alpha_2 x_2 + \cdots + \alpha_{n-1} x_{n-1} + x_n,$$

so that

$$\|\alpha_1 x_1 + \alpha_2 x_2 + \cdots + \alpha_{n-1} x_{n-1} + x_n\| \geq c_n,$$

from which it follows that for any scalar $\alpha_n$ with $\alpha_n \neq 0$,

$$\left\| \frac{\alpha_1}{\alpha_n} x_1 + \frac{\alpha_2}{\alpha_n} x_2 + \cdots + \frac{\alpha_{n-1}}{\alpha_n} x_{n-1} + x_n \right\| \geq c_n,$$

$$i.e., \|\alpha_1 x_1 + \alpha_2 x_2 + \cdots + \alpha_{n-1} x_{n-1} + \alpha_n x_n\| \geq |\alpha_n| c_n.$$

Instead of

$$M = [\{x_1, x_2, \ldots, x_{n-1}\}],$$

we could have taken

$$[\{x_1, x_2, \ldots, x_{i-1}, x_{i+1}, \ldots, x_n\}],$$

whose dimension is $(n-1)$ and we would have arrived at

$$\|\alpha_1 x_1 + \alpha_2 x_2 + \cdots + \alpha_n x_n\| \geq |\alpha_i| c_i,$$

for every $i = 1, 2, \ldots, n$. Consequently,

$$\|\alpha_1 x_1 + \alpha_2 x_2 + \cdots + \alpha_n x_n\| \geq a \|x\|_0,$$

where $a = \min_{1 \leq i \leq n} c_i > 0$,

$$i.e., \|x\| \geq a \|x\|_0,$$

*i.e.*, (2.3) holds.

The proof of the theorem is now complete. □

**Theorem 16.** *If $X$ is a finite-dimensional normed linear space, $\tilde{X}$ is finite dimensional too and dim $\tilde{X} =$ dim $X$.*

*Proof.* Let $X$ be a finite-dimensional normed linear space and let $\{x_1, x_2, \ldots, x_n\}$ be a basis of $X$. Let $x \in X$. Then there exist unique scalars $\alpha_1, \alpha_2, \ldots, \alpha_n$, associated with $x$, such that

$$x = \sum_{i=1}^{n} \alpha_i x_i.$$

For $i = 1, 2, \ldots, n$, define the functional $f_i$ by

$$f_i(x) = \alpha_i, \text{ for all } x \in X.$$

Note that $f_i$ is a linear functional on $X$, $i = 1, 2, \ldots, n$. We now claim that the set $\{f_1, f_2, \ldots, f_n\}$ is linearly independent. If $\sum_{i=1}^{n} \beta_i f_i = 0$, where $\beta_i$, $i = 1, 2, \ldots, n$ are scalars, then for $j = 1, 2, \ldots, n$,

$$\begin{aligned} 0 = \left( \sum_{i=1}^{n} \beta_i f_i \right)(x_j) &= \sum_{i=1}^{n} \beta_i f_i(x_j) \\ &= \beta_j, \text{ since } f_i(x_i) = 1 \text{ and} \\ &\quad f_i(x_j) = 0, \text{if } i \neq j, \end{aligned}$$

which implies that

$$\beta_j = 0, j = 1, 2, \ldots, n,$$

proving our claim. Now, if $f$ is any linear functional and $x = \sum_{i=1}^{n} \alpha_i x_i$, then

$$\begin{aligned} f(x) &= f\left(\sum_{i=1}^{n} \alpha_i x_i\right) \\ &= \sum_{i=1}^{n} \alpha_i f(x_i) \\ &= \sum_{i=1}^{n} f_i(x) f(x_i) \\ &= \sum_{i=1}^{n} \gamma_i f_i(x), \text{ where } \gamma_i = f(x_i), i = 1, 2, \dots, n \\ &= \left(\sum_{i=1}^{n} \gamma_i f_i\right)(x), \text{ for all } x \in X, \end{aligned}$$

from which it follows that

$$f = \sum_{i=1}^{n} \gamma_i f_i.$$

Thus, $\{f_1, f_2, \dots, f_n\}$ is a basis of $\tilde{X}$ and so

$$dim\ \tilde{X} = dim\ X = n,$$

completing the proof. □

---

## 2.4 Convexity

In this section, we introduce the notion of "convexity." Throughout, we shall suppose that $X$ is a real or complex linear space.

**Definition 28.** *The segment connecting the vectors $x, y \in X$ is the set of all vectors*

$$\{(1-\alpha)x + \alpha y / 0 \leq \alpha \leq 1\}.$$

*Equivalently, it is the set of all vectors*

$$\{\alpha x + \beta y / \alpha + \beta = 1, \alpha, \beta \geq 0\}.$$

**Definition 29.** *A subset $S$ of $X$ is said to be "convex," if whenever $x, y \in S$, the segment connecting them is contained in $S$.*

Note that the concept of a segment in the Euclidean plane is the usual "line segment."

**Theorem 17.** *Let $S$ be convex and let $x_1, x_2, \ldots, x_n \in S$. Then all elements of the form*

$$\alpha_1 x_1 + \alpha_2 x_2 + \cdots + \alpha_n x_n,$$

*where $\alpha_i \geq 0$, $i = 1, 2, \ldots, n$ and $\alpha_1 + \alpha_2 + \cdots + \alpha_n = 1$, also belong to $S$.*

*Proof.* The proof is by induction on "$n$." The result is true when $n = 2$, by the definition of a convex set. We shall suppose that the result holds for $(n-1)$. Let us prove that the result holds for $n$. Let $\alpha_i \geq 0$, $i = 1, 2, \ldots, n$ be such that $\alpha_1 + \alpha_2 + \cdots + \alpha_n = 1$.
Case 1. If $\alpha_1 + \alpha_2 + \cdots + \alpha_{n-1} = 0$, it is clear that $\alpha_i = 0$, for $i = 1, 2, \ldots, (n-1)$ and the result holds trivially in this case.
Case 2. Suppose $\beta = \alpha_1 + \alpha_2 + \cdots + \alpha_{n-1} > 0$.
Then, $\beta + \alpha_n = 1$. By induction assumption,

$$\frac{\alpha_1}{\beta} x_1 + \frac{\alpha_2}{\beta} x_2 + \cdots + \frac{\alpha_{n-1}}{\beta} x_{n-1} \in S.$$

Since $S$ is convex and $\beta + \alpha_n = 1$, it follows that

$$\beta\left(\frac{\alpha_1}{\beta} x_1 + \frac{\alpha_2}{\beta} x_2 + \cdots + \frac{\alpha_{n-1}}{\beta} x_{n-1}\right) + \alpha_n x_n \in S,$$

$$\text{i.e., } \alpha_1 x_1 + \alpha_2 x_2 + \cdots + \alpha_{n-1} x_{n-1} + \alpha_n x_n \in S,$$

completing the proof. □

**Theorem 18.** *Let $S$ be a convex subset of a normed linear space $X$. Then $\overline{S}$ is also convex.*

*Proof.* Let $x, y \in \overline{S}$ and $\epsilon > 0$. We can find $x_1, y_1 \in S$ such that

$$\|x - x_1\| < \epsilon \text{ and } \|y - y_1\| < \epsilon.$$

Let $\alpha, \beta \geq 0$ such that $\alpha + \beta = 1$. Then

$$\begin{aligned}
&\|(\alpha x + \beta y) - (\alpha x_1 + \beta y_1)\| \\
&\quad = \|\alpha(x - x_1) + \beta(y - y_1)\| \\
&\quad \leq \|\alpha(x - x_1)\| + \|\beta(y - y_1)\| \\
&\quad = \alpha\|x - x_1\| + \beta\|y - y_1\| \\
&\quad < (\alpha + \beta)\epsilon \\
&\quad = \epsilon, \text{ since } \alpha + \beta = 1.
\end{aligned}$$

Note that $\alpha x_1 + \beta y_1 \in S$, since $S$ is convex. It now follows from the above that

$$\alpha x + \beta y \in \overline{S},$$

$$i.e., \overline{S} \text{ is convex.}$$

□

It is clear that the intersection of any collection of convex sets of a linear space $X$ is also convex.

**Definition 30.** *Let $S$ be a subset of a linear space $X$. Then the intersection of all convex sets of $X$ containing $S$ is called the "convex hull" of $S$, denoted by $S_c$.*

In the context of Definition 30, it is worthwhile to note that there always exists a convex set containing $S$, namely $X$ itself. We also note that every subspace of $X$ is a convex set.

**Theorem 19.** *Let $S$ be a subset of a linear space $X$. The convex hull of $S$ consists of all vectors of the form*

$$\alpha_1 x_1 + \alpha_2 x_2 + \cdots + \alpha_{n-1} x_{n-1} + \alpha_n x_n, \text{ where } x_i \in S, \alpha_i \geq 0, \text{ for } i = 1, 2, \ldots, n$$

*and* $\sum_{i=1}^{n} \alpha_i = 1$, *noting that the index $n$ is not fixed.*

*Proof.* Let $M$ be the set of all vectors of the form given in the statement of the theorem. Note that $M$ is a convex set containing $S$. It now follows that $M$ contains $S_c$, the convex hull of $S$. In view of Theorem 17, any convex set containing $S$ must also contain $M$, i.e., $S_c$ should contain $M$ and so $S_c = M$, completing the proof. □

**Definition 31.** *Let $S$ be a subset of a normed linear space $X$. The closure $\overline{S}_c$ of the convex hull $S_c$ of $S$ is called the "closed convex hull" of $S$.*

**Theorem 20.** *Let*

$$E = \bigcap_M \{M/M \supset S, M \text{ convex and closed}\},$$

*i.e., $E$ is the intersection of all closed, convex sets containing $S$. Then*

$$E = \overline{S}_c.$$

*Proof.* Since $S_c$ is convex, $\overline{S}_c$ is convex. $\overline{S}_c$ is already closed and $\overline{S}_c \supset S$. Thus

$$E \subset \overline{S}_c.$$

Noting that $S_c$ is the intersection of all convex sets containing $S$, while $E$ is the intersection of all closed, convex sets containing $S$,

$$E \supset S_c.$$

This implies that

$$\overline{E} \supset \overline{S}_c,$$
$$i.e., E \supset \overline{S}_c, \text{ since } E = \overline{E}, E \text{ being closed.}$$

Consequently,

$$E = \overline{S}_c,$$

completing the proof. □

**Exercises**

1. Let

$$f_1(x) = \lim_{n\to\infty} x_n, x = \{x_n\} \in c;$$

$$f_2(x) = \sum_{n=1}^{\infty} \frac{x_n}{n^2}, x = \{x_n\} \in \ell_\infty,$$

where, as usual, $c, \ell_\infty$ denote the spaces of convergent and bounded sequences respectively. Prove that $f_1$ and $f_2$ are bounded linear functionals on $c, \ell_\infty$ respectively. Find their norms too.

2. Let $X, Y$ be normed linear spaces and $f : X \to Y$ be linear. Prove that the following statements are equivalent:

   (i) $f$ is continuous;

   (ii) $f$ maps Cauchy sequences in $X$ to Cauchy sequences in $Y$;

   and

   (iii) $f$ maps convergent sequences in $X$ to convergent sequences in $Y$.

3. Let $\|.\|_1, \|.\|_2$ be equivalent norms on a linear space $X$. Prove the following:

   (i) $\{x_n\}$ is Cauchy with respect to $\|.\|_1$ if and only if $\{x_n\}$ is Cauchy with respect to $\|.\|_2$;

   (ii) $x_n \to x$, $n \to \infty$ with respect to $\|.\|_1$ if and only if $x_n \to x$, $n \to \infty$ with respect to $\|.\|_2$;

   and

   (iii) $E \subset X$ is open with respect to $\|.\|_1$ if and only if $E$ is open with respect to $\|.\|_2$.

4. Let $X_1, X_2, \ldots, X_n$ be normed linear spaces. Prove that the following are equivalent norms on $X_1 \times X_2 \times \cdots \times X_n$.

$$\|(x_1, x_2, \ldots, x_n)\|_1 = \sum_{i=1}^{n} \|x_i\|, x_i \in X_i, i = 1, 2, \ldots, n;$$

$$\|(x_1, x_2, \ldots, x_n)\|_2 = \left(\sum_{i=1}^{n} \|x_i\|^2\right)^{\frac{1}{2}}, x_i \in X_i, i = 1, 2, \ldots, n;$$

$$\|(x_1, x_2, \ldots, x_n)\|_3 = \max_{1 \le i \le n} \|x_i\|, x_i \in X_i, i = 1, 2, \ldots, n.$$

5. Prove that two norms on a linear space $X$ are equivalent if and only if every Cauchy sequence with respect to one of the norms is a Cauchy sequence with respect to the other norm.

6. Let $X, Y$ be normed linear spaces and $X$ be finite-dimensional. Prove that any linear mapping from $X$ to $Y$ is continuous.

7. Prove that any two norms on a finite-dimensional linear space $X$ are equivalent.

8. If $X$ is any finite-dimensional normed linear space, prove that $X$ is a Banach space.

9. If $M$ is any finite-dimensional subspace of a normed linear space $X$, prove that $M$ is closed.

10. Let $X, Y$ be normed linear spaces with $X \neq \{0\}$. If $B(X, Y)$ is a Banach space, prove that $Y$ is a Banach space.

11. If the normed linear spaces $X, Y$ are linearly homeomorphic and if one of them is a Banach space, prove that the other one is also a Banach space. Deduce that all finite-dimensional normed linear spaces are Banach spaces.

12. Let $e = \{1, 1, 1, \ldots\}$ and $e_i = \{0, 0, \ldots, 0, 1, 0, \ldots\}, 1$ occurring in the $i$th place, $i = 1, 2, \ldots$. Prove that the set $\{e, e_1, e_2, \ldots\}$ is a basis of $c$, the space of all convergent sequences.

# Chapter 3

# Hahn–Banach Theorem

## 3.1 Hahn–Banach theorem for the real normed linear space

**Definition 32.** *Let $X$ be a linear space. Let*

$$p : X \to \mathbb{R}$$

*be such that*

*(i)* $p(x) \geq 0$*, for all* $x \in X$*;*

*(ii)* $p(x + y) \leq p(x) + p(y)$*, for all* $x, y \in X$

*(subadditive);*

*and*

*(iii)* $p(\alpha x) = \alpha p(x)$*, for all* $\alpha \geq 0$ *and* $x \in X$

*(positive homogeneous).*

*Such a mapping $p$ is called a "convex functional." A mapping that is subadditive and positive homogeneous is called a "sublinear functional."*

Note that every norm is a convex functional.

**Lemma 3.** *Let $M$ be a proper subspace of the real linear space $X$. Let $x_0 \in X - M$. Consider the subspace of $X$ spanned by $M$ and $\{x_0\}$, i.e.,*

$$N = [M \cup \{x_0\}].$$

*Let $f$ be a linear functional defined on $M$. Let $p$ be a sublinear functional defined on $X$. Let*

$$f(x) \leq p(x), \text{ for all } x \in M.$$

*Then $f$ can be extended to a linear functional $F$ on $N$ such that*

$$F(x) \leq p(x), \text{ for all } x \in N.$$

*Proof.* Since $f(x) \leq p(x)$, for all $x \in M$, for $y_1, y_2 \in M$,

$$\begin{aligned} f(y_1 - y_2) &\leq p(y_1 - y_2), \\ \text{i.e., } f(y_1) - f(y_2) &\leq p(y_1 - y_2) \\ &= p(y_1 + x_0 - y_2 - x_0) \\ &\leq p(y_1 + x_0) + p(-y_2 - x_0), \\ \text{i.e., } -p(-y_2 - x_0) - f(y_2) &\leq p(y_1 + x_0) - f(y_1). \end{aligned} \tag{3.1}$$

Keeping $y_1$ fixed and allowing $y_2$ to vary through $M$ in (3.1), the set of numbers

$$\{-p(-y_2 - x_0) - f(y_2)/y_2 \in M\}$$

has an upper bound and hence a least upper bound. Let

$$a = \sup_{y_2 \in M} \{-p(-y_2 - x_0) - f(y_2)\}.$$

Similarly, let

$$b = \inf_{y_1 \in M} \{p(y_1 + x_0) - f(y_1)\},$$

using (3.1) again. In view of (3.1), we have

$$a \leq b.$$

Then we can find a real number $c_0$ such that

$$a \leq c_0 \leq b.$$

Now, for any $y \in M$, we have

$$-p(-y - x_0) - f(y) \leq a \leq c_0 \leq b \leq p(y + x_0) - f(y). \tag{3.2}$$

Since $x_0 \notin M$, any $x \in N$ can be written as

$$x = y + \alpha x_0,$$

where $\alpha$ is a unique scalar and $y$ is a unique vector in $M$. Because of the above unique representation, the mapping

$$F : N \to \mathbb{R}$$

defined by

$$F(x) = F(y + \alpha x_0) = f(y) + \alpha c_0,$$

is well defined. The reader can easily verify that $F$ is a linear functional on $N$. Note that if $y \in M$, $F(y) = f(y)$, so that $F$ extends $f$. It remains to prove that

$$F(x) \leq p(x), \text{ for all } x \in N.$$

Now, for any $x \in N$, $x = y + \alpha x_0$, where one of $\alpha = 0$, $\alpha > 0$, $\alpha < 0$ must hold.

**Case 1.** If $\alpha = 0$, $F(x) = F(y + \alpha x_0) = F(y) = f(y)$ and we are through in this case.

**Case 2.** Let $\alpha > 0$. In the right half of (3.2), replace $y$ by $\frac{y}{\alpha}$ so that

$$c_0 \leq p\left(\frac{y}{\alpha} + x_0\right) - f\left(\frac{y}{\alpha}\right),$$

so that

$$f(y) + \alpha c_0 \leq p(y + \alpha x_0),$$

$$i.e., F(x) \leq p(x).$$

**Case 3.** Let $\alpha < 0$. Replace $y$ by $\frac{y}{\alpha}$ in the left half of (3.2) to get

$$-p\left(-\frac{y}{\alpha} - x_0\right) - f\left(\frac{y}{\alpha}\right) \leq c_0,$$

$$i.e., -p\left(-\frac{y}{\alpha} - x_0\right) \leq c_0 + f\left(\frac{y}{\alpha}\right),$$

$$i.e., (-\alpha)p\left(-\frac{y}{\alpha} - x_0\right) \geq \alpha c_0 + f(y), \text{ since } \alpha < 0,$$

$$i.e., p(y + \alpha x_0) \geq \alpha c_0 + f(y),$$

$$i.e., F(x) \leq p(x),$$

completing the proof of the lemma. □

We need to introduce some more concepts before we take up the Hahn–Banach theorem.

**Definition 33.** *Let $R$ be a relation in a set $X$. $X$ is said to be "partially ordered" under $R$ if the following are satisfied:*

*(i) $aRa$, for all $a \in X$;*

*(ii) $aRb$ and $bRc$ imply $aRc$, for all $a, b, c \in X$;*

*and*

*(iii) if $aRb$ and $bRa$, then $a = b$, for all $a, b \in X$.*

*Further, if for any two elements $a, b \in X$, one of $aRb$ or $bRa$ holds, then $X$ is said to be "totally ordered" under $R$. A totally ordered set is called a "chain."*

**Definition 34.** *Let $X$ be partially ordered under the relation $R$ and $A$ be a subset of $X$. The element $a \in X$ is called an "upper bound" for $A$ if $yRa$, for all $y \in A$.*

**Definition 35.** *Let $X$ be partially ordered under $R$. The element $a \in X$ is called a "maximal element" of $X$ if $aRy$ implies that $a = y$, for any $y \in X$.*

We note that we do not require a maximal element to be the biggest element in a given set; we only require that nothing is bigger than it. We further note that there is nothing unique about the upper bound or the maximal element. There may be many of each of them.

**Definition 36.** *The partially ordered set $X$, under the relation $R$, is said to be "inductively ordered" if every totally ordered subset of $X$ has an upper bound in $X$.*

We further need the following:

**Lemma 4** (Zorn's lemma). *Every non-empty inductively ordered set in $X$ has a maximal element.*

**Theorem 21** (Hahn–Banach). *Let $M$ be a subspace of the linear space $X$. Let $p$ be a sublinear functional on $X$. Let $f$ be a linear functional on $M$ such that*

$$f(x) \le p(x), \text{ for all } x \in M.$$

*Then there exists a linear functional $F$ on $X$, extending $f$, such that*

$$F(x) \le p(x), \text{ for all } x \in X.$$

*Proof.* Let $S$ be the set of all linear functionals $\tilde{f}$, extending $f$, such that

$$\tilde{f}(x) \le p(x), \text{ for all } x \in \mathcal{D}(\tilde{f}),$$

where $\mathcal{D}(\tilde{f})$ is the domain of $\tilde{f}$. Note that this set is non-empty, since $f$ itself is a member of $S$. We also note that, in view of Lemma 3, $S$ contains non-trivial elements too. For $\tilde{f}_1, \tilde{f}_2 \in S$, define

$$\tilde{f}_1 < \tilde{f}_2$$

if and only if $\tilde{f}_2$ extends $\tilde{f}_1$ or equivalently, $\mathcal{D}(\tilde{f}_1) \subset \mathcal{D}(\tilde{f}_2)$ and $\tilde{f}_2/\mathcal{D}(\tilde{f}_1) = \tilde{f}_1$. The reader can easily check that $<$ is a partial ordering on $S$. We now show that $S$ is inductively ordered under the relation $<$. For this purpose, let $T = \{\tilde{f}_\alpha\}$ be a chain in $S$. We have to prove the existence of $\hat{f} \in S$, which is an upper bound of $T$. Consider the function $\hat{f}$, whose domain is $\bigcup_\alpha \mathcal{D}(\tilde{f}_\alpha)$, defined as follows: For $x \in \bigcup_\alpha \mathcal{D}(\tilde{f}_\alpha)$, there exists an $\alpha$ such that $x \in \mathcal{D}(\tilde{f}_\alpha)$ and we define

$$\hat{f}(x) = \tilde{f}_\alpha(x).$$

We first claim that $\bigcup_\alpha \mathcal{D}(\tilde{f}_\alpha)$ is a subspace. Let $x, y \in \bigcup_\alpha \mathcal{D}(\tilde{f}_\alpha)$. Then there exist $\alpha_1, \alpha_2$ such that $x \in \mathcal{D}(\tilde{f}_{\alpha_1})$, $y \in \mathcal{D}(\tilde{f}_{\alpha_2})$. Since $T$ is totally ordered, either $\mathcal{D}(\tilde{f}_{\alpha_1}) \subset \mathcal{D}(\tilde{f}_{\alpha_2})$ or vice versa. Let $\mathcal{D}(\tilde{f}_{\alpha_1}) \subset \mathcal{D}(\tilde{f}_{\alpha_2})$, noting that the other case can be dealt with in a similar fashion. We now have $x, y \in \mathcal{D}(\tilde{f}_{\alpha_2})$. Since $\mathcal{D}(\tilde{f}_{\alpha_2})$ is a subspace, $\beta_1 x + \beta_2 y \in \mathcal{D}(\tilde{f}_{\alpha_2})$ and so

$$\beta_1 x + \beta_2 y \in \bigcup_\alpha \mathcal{D}(\tilde{f}_\alpha), \text{ for all scalar } \beta_1, \beta_2,$$

which establishes our claim.

Next, we claim that $\hat{f}$ is well defined. Suppose $x \in \mathcal{D}(\tilde{f}_\alpha)$ and $x \in \mathcal{D}(\tilde{f}_\beta)$. Using the definition of $\hat{f}$, $\hat{f}(x) = \tilde{f}_\alpha(x)$ and $\hat{f}(x) = \tilde{f}_\beta(x)$. In view of the total ordering of $T$, either $\tilde{f}_\alpha$ extends $\tilde{f}_\beta$ or the converse is true. Whatever the case,

$$\tilde{f}_\alpha(x) = \tilde{f}_\beta(x),$$

so that $\hat{f}$ is well defined, proving the claim. It is clear that $\hat{f}$ is a linear mapping extending $f$, $\hat{f}(x) \leq p(x)$, for all $x \in \mathcal{D}(\hat{f})$ and for any $\tilde{f}_\alpha \in T$, $\tilde{f}_\alpha < \hat{f}$. Thus, $\hat{f}$ is an upper bound of $T$ in $S$ and so $S$ is inductively ordered. Now, applying Zorn's lemma, there exists a maximal element $F$ in $S$. Since $F \in S$, $F$ is a linear functional extending $f$ such that

$$F(x) \leq p(x), \text{ for all } x \in \mathcal{D}(F).$$

To complete the proof, it remains to prove that

$$\mathcal{D}(F) = X.$$

Suppose not. Then there exists $x_0 \in X$ such that $x_0 \notin \mathcal{D}(F)$. Applying Lemma 3, $F$ can be extended to a linear functional $F'$ extending $f$ such that

$$F'(x) \leq p(x), \text{ for all } x \in [\mathcal{D}(F) \cup \{x_0\}].$$

Thus, $F' \in S$, which is a contradiction of the maximality of $F$, since $F'$ extends $F$. In other words, no such $x_0$ exists and consequently $\mathcal{D}(F) = X$, completing the proof of the theorem. □

---

## 3.2 Hahn–Banach theorem for the complex normed linear space

**Definition 37.** *The convex functional $p(x)$ is said to be "symmetric," if for all scalars $\alpha$, $p(\alpha x) = |\alpha| p(x)$.*

**Theorem 22** (Hahn–Banach). *Let $M$ be a subspace of the complex linear space $X$. Let $p$ be a symmetric convex functional defined on $X$. Let $f$ be a linear functional defined on $M$ such that*

$$|f(x)| \le p(x), \text{ for all } x \in M.$$

*Then there exists a linear functional $F$ defined on $X$, extending $f$, such that*

$$|F(x)| \le p(x), \text{ for all } x \in X.$$

*Proof.* We first note that the complex linear space $X$ can be viewed as a real linear space, by restricting the scalars to real numbers. Let

$$f(x) = f_1(x) + if_2(x), x \in X.$$

We now claim that $f_1$ and $f_2$ are real-valued, real linear functionals, where, by a real linear functional on the complex linear space $X$, we mean a functional $g$ such that

$$g(\alpha x) = \alpha g(x), \text{ for } x \in X \text{ and } \alpha \in \mathbb{R}.$$

Let $\alpha \in \mathbb{R}$. Now,

$$f(\alpha x) = \alpha f(x),$$
$$i.e., f_1(\alpha x) + if_2(\alpha x) = \alpha[f_1(x) + if_2(x)].$$

Equating the real and imaginary parts, we have

$$f_1(\alpha x) = \alpha f_1(x)$$

and

$$f_2(\alpha x) = \alpha f_2(x).$$

In a similar fashion, we can prove that sums are also preserved. It then follows that $f_1, f_2$ are real linear functionals. Now,

$$\begin{aligned} i[f_1(x) + if_2(x)] &= if(x) \\ &= f(ix) \\ &= f_1(ix) + if_2(ix), \end{aligned}$$

from which we have

$$f_1(ix) = -f_2(x). \tag{3.3}$$

It is clear that

$$f_1(x) \leq p(x), \text{ for all } x \in M.$$

Hence, by Theorem 21, there exists a real-valued, real linear functional $F_1$ on $X$, extending $f_1$, such that

$$F_1(x) \leq p(x), \text{ for all } x \in X.$$

Now we define

$$F(x) = F_1(x) - iF_1(ix), \text{ for } x \in X.$$

We claim that $F$ extends $f$. Let $x \in M$. Since $F_1$ extends $f_1$,

$$F_1(x) = f_1(x)$$

and

$$\begin{aligned} F_1(ix) &= f_1(ix) \\ &= -f_2(x), \text{ using (3.3)}, \end{aligned}$$

so that

$$\begin{aligned} F(x) &= f_1(x) + if_2(x) \\ &= f(x). \end{aligned}$$

It is clear that $F$ is a real linear functional, proving our claim. We next show that $F$ is a complex, linear functional. It suffices to prove that

$$F(ix) = iF(x).$$

Now,

$$\begin{aligned} F(ix) &= F_1(ix) - iF_1(i^2x) \\ &= F_1(ix) - iF_1(-x) \\ &= F_1(ix) + iF_1(x); \end{aligned}$$

$$\begin{aligned} iF(x) &= i[F_1(x) - iF_1(ix)] \\ &= F_1(ix) + iF_1(x), \end{aligned}$$

so that

$$F(ix) = iF(x).$$

Finally, we claim that

$$|F(x)| \leq p(x), \text{ for all } x \in X.$$

To this end, we use the fact that $p$ is a symmetric, convex functional. The inequality trivially holds if $F(x) = 0$. Let $F(x) \neq 0$. Let $F(x) = re^{i\theta}$. Then

$$F(e^{-i\theta}x) = r = |F(x)|$$

and so $F(e^{-i\theta}x)$ is purely real. Thus, $-F_1(ie^{-i\theta}x) = 0$ and

$$F(e^{-i\theta}x) = F_1(e^{-i\theta}x).$$

Consequently,

$$\begin{aligned} |F(x)| &= F(e^{-i\theta}x) \\ &= F_1(e^{-i\theta}x) \\ &= |F_1(e^{-i\theta}x)|, \text{ since } |F(x)| \geq 0 \\ &\leq p(e^{-i\theta}x) \\ &= p(x), \text{ for all } x \in X, \end{aligned}$$

completing the proof of the theorem. □

---

## 3.3 Some consequences of the Hahn–Banach theorem

We now present some important consequences of the Hahn–Banach theorem.

**Theorem 23.** *Let $M$ be a subspace of the complex normed linear space $X$. Let $f$ be a bounded linear functional on $M$. Then there exists a bounded linear functional $F$ on $X$, extending $f$, such that $\|F\| = \|f\|$.*

*Proof.* Since $f$ is a bounded linear functional on $M$,

$$|f(x)| \leq \|f\|\|x\|, \text{ for all } x \in M.$$

Now, for every $x \in X$, we define

$$p(x) = \|f\|\|x\|.$$

The reader can easily prove that $p$ is a symmetric, convex functional. We also have

$$|f(x)| \leq p(x), \text{ for all } x \in M.$$

By Theorem 22, we can extend $f$ to a linear functional $F$ on $X$ such that

$$|F(x)| \leq p(x) = \|f\|\|x\|, \text{ for all } x \in X.$$

It is now clear that $F$ is a bounded linear functional on $X$ such that

$$\|F\| \leq \|f\|. \tag{3.4}$$

Now, for $x \in M$,

$$|f(x)| = |F(x)| \leq \|F\|\|x\|,$$

which implies that

$$\|f\| \leq \|F\|. \tag{3.5}$$

From (3.4) and (3.5), we get

$$\|F\| = \|f\|,$$

completing the proof of the theorem. □

Note that Theorem 23, with suitable modification, holds for real linear spaces too.

**Theorem 24.** *Let $X$ be a normed linear space, $x_0 \in X$, $x_0 \neq 0$. Then there exists a bounded linear functional $F$, defined on $X$, such that*

$$\|F\| = 1 \text{ and } F(x_0) = \|x_0\|.$$

*Proof.* We shall consider the subspace

$$M = [\{x_0\}],$$

consisting of all scalar multiples of $x_0$. Consider the functional $f$ on $M$ defined by

$$f : M \to \mathbb{C}(\text{or } \mathbb{R})$$
$$\alpha x_0 \to \alpha\|x_0\|.$$

It is clear that $f$ is a linear functional and $f(x_0) = \|x_0\|$.
For any $x \in M$,

$$|f(x)| = |\alpha|\|x_0\| = \|\alpha x_0\| = \|x\|. \tag{3.6}$$

From (3.6), we note that $f$ is a bounded linear functional and $\|f\| \leq 1$. If $\|f\| < 1$, then

$$|f(x)| \leq \|f\|\|x\| < \|x\|,$$

which contradicts (3.6), so that

$$\|f\| = 1.$$

In view of Theorem 23, $f$ can be extended to a bounded linear functional $F$ defined on $X$ such that

$$\|F\| = \|f\| = 1.$$

Also,

$$F(x_0) = f(x_0) = \|x_0\|,$$

completing the proof. □

**Theorem 25.** *Let $M$ be a subspace of the normed linear space $X$. For $x_1 \in X$, $x_1 \notin M$, let the distance from $x_1$ to $M$, i.e., $d(x_1, M) = d > 0$. Then there exists a bounded linear functional on $X$ such that*

$$\|F\| = 1, F(x_1) = d \text{ and } F(x) = 0, \text{ for all } x \in M.$$

*Proof.* Let

$$N = [M \cup \{x_1\}].$$

Then, any $y \in N$ can be uniquely written as

$$y = x + \alpha x_1,$$

where $x \in M$ and $\alpha$ is a scalar. Consider the functional

$$f : N \to \mathbb{C}(\text{or } \mathbb{R})$$
$$y \to \alpha d,$$

where $y$ is as mentioned above and $d$ is as in the hypothesis. It is clear that $f$ is well defined and it is also linear. Further, if $x \in M$,

$$f(x) = 0 \text{ and } f(x_1) = d.$$

We now claim that $f$ is bounded and $\|f\| = 1$.
Let $\alpha \neq 0$ be a scalar. Now,

$$\begin{aligned} \|y\| &= \|x + \alpha x_1\| \\ &= \left\|-\alpha\left(-\frac{x}{\alpha} - x_1\right)\right\| \\ &= |\alpha| \left\|-\frac{x}{\alpha} - x_1\right\|. \end{aligned}$$

Since

$$d = d(x_1, M) = \inf_{x \in M} \|x - x_1\|$$

and $-\frac{x}{\alpha} \in M$, we have

$$\left\|-\frac{x}{\alpha} - x_1\right\| \geq d,$$

so that

$$\|y\| \geq |\alpha| d,$$

$$i.e., |f(y)| = |\alpha| d \leq \|y\|, \text{ for all } y \in N.$$

This implies that $f$ is a bounded linear functional and

$$\|f\| \leq 1. \tag{3.7}$$

Now, for any $\epsilon > 0$, there exists $x \in M$ such that

$$\|x - x_1\| < d + \epsilon. \tag{3.8}$$

Let $z = \frac{x - x_1}{\|x - x_1\|}$. Then $\|z\| = 1$. Also,

$$\begin{aligned} |f(z)| &= \frac{1}{\|x - x_1\|} |f(x - x_1)| \\ &= \frac{d}{\|x - x_1\|}, \text{ using the definition of } f \\ &> \frac{d}{d + \epsilon}, \text{ in view of (3.8).} \end{aligned}$$

We know that

$$\|f\| = \sup_{\|y\|=1} |f(y)|.$$

Since $\|z\| = 1$,

$$\|f\| > |f(z)|,$$

so that

$$\|f\| > \frac{d}{d + \epsilon}.$$

Since $\epsilon$ is arbitrary, it follows that

$$\|f\| \geq 1. \tag{3.9}$$

In view of (3.7) and (3.9), we have

$$\|f\| = 1.$$

Hence there exists a bounded linear functional $F$ on $X$ such that

$$\|F\| = \|f\| = 1.$$

The proof of the theorem is now complete. □

**Definition 38.** *A non-empty subset $S$ of a normed linear space $X$ is said to be "fundamental" in $X$, if the linear hull of $S$ is dense in $X$.*

**Theorem 26.** *A normed linear space $X$ is separable if and only if there exists a countable subset $S$ of $X$ that is fundamental in $X$.*

*Proof.* Let $X$ be separable and $S$ be a countable subset of $X$ that is dense in $X$. Then the linear hull of $S$, which contains $S$, must be dense in $X$ and so $S$ is fundamental in $X$.

Conversely, suppose there exists a countable subset $S$ of $X$ that is fundamental in $X$. Let $K_0$ be a subset of the scalar field $K$ that is dense in $K$, where $K = \mathbb{R}$ or $\mathbb{C}$. For instance, if $K = \mathbb{R}$, choose $K_0$ to be the set of all rational numbers; if $K = \mathbb{C}$, choose $K_0$ to be the set of all complex numbers with rational real and imaginary parts. Let $\beta$ be the set of all finite linear combinations of the form

$$\beta_1 x_1 + \beta_2 x_2 + \cdots + \beta_n x_n,$$

where $\beta_i \in K_0$, $x_i \in S$, $i = 1, 2, \ldots, n$, $n$ being an arbitrary positive integer. We first claim that $\beta$ is countable. Since $K_0$ and $S$ are countable, $K_0 \times S$ is countable. Also, the family $\mathcal{A}$ of all finite subsets of $K_0 \times S$ is countable. The mapping

$$\{(\beta_1, x_1), (\beta_2, x_2), \ldots, (\beta_n x_n)\} \to \beta_1 x_1 + \beta_2 x_2 + \cdots + \beta_n x_n$$

maps $\mathcal{A}$ onto $\beta$ and so $\beta$ is countable, proving the claim. We now show that $\beta$ is dense in $X$. Let $x \in X$ and $\epsilon > 0$. Since the linear hull of $S$ is dense in $X$, there exist $x_1, x_2, \ldots, x_n \in S$ and $\alpha_1, \alpha_2, \ldots, \alpha_n \in K$ such that

$$\|x - (\alpha_1 x_1 + \alpha_2 x_2 + \cdots + \alpha_n x_n)\| < \frac{\epsilon}{2}. \tag{3.10}$$

Again, since $K_0$ is dense in $K$, there exist $\beta_1, \beta_2, \ldots, \beta_n \in K_0$ such that

$$|\alpha_i - \beta_i| < \frac{\epsilon}{2n\|x_i\|}, \quad \text{for } i = 1, 2, \ldots, n. \tag{3.11}$$

Then,

$$\beta_1 x_1 + \beta_2 x_2 + \cdots + \beta_n x_n \in \beta$$

and

$$\begin{aligned}
&\|x - (\beta_1 x_1 + \beta_2 x_2 + \cdots + \beta_n x_n)\| \\
&\quad = \|\{x - (\alpha_1 x_1 + \alpha_2 x_2 + \cdots + \alpha_n x_n)\} \\
&\qquad + \{(\alpha_1 - \beta_1)x_1 + (\alpha_2 - \beta_2)x_2 + \cdots + (\alpha_n - \beta_n)x_n\}\| \\
&\quad \leq \|\{x - (\alpha_1 x_1 + \alpha_2 x_2 + \cdots + \alpha_n x_n)\| \\
&\qquad + \|(\alpha_1 - \beta_1)x_1 + (\alpha_2 - \beta_2)x_2 + \cdots + (\alpha_n - \beta_n)x_n\| \\
&\quad < \frac{\epsilon}{2} + |\alpha_1 - \beta_1|\|x_1\| + |\alpha_2 - \beta_2|\|x_2\| + \cdots \\
&\qquad + |\alpha_n - \beta_n|\|x_n\|, \quad \text{using (3.10)} \\
&\quad < \frac{\epsilon}{2} + \frac{\epsilon}{2}, \quad \text{using (3.11)} \\
&\quad = \epsilon,
\end{aligned}$$

from which it follows that $\beta$ is dense in $X$. The proof of the theorem is now complete. □

**Theorem 27.** *If $\tilde{X}$ is separable, so is $X$, where $\tilde{X}$ denotes the dual space of the normed linear space $X$.*

*Proof.* Consider the set

$$S = \{f \in \tilde{X} / \|f\| = 1\}.$$

Since any subset of a separable metric space is separable, $S$ is separable. Let $\{f_1, f_2, \ldots\}$ be a countable dense subset of $S$. $\|f_n\| = 1$, $n = 1, 2, \ldots$, since $f_n \in S$, $n = 1, 2, \ldots$. Since

$$\|f_n\| = \sup_{\|x\|=1} |f_n(x)|,$$

there exists $x_n$ with $\|x_n\| = 1$ such that $|f_n(x_n)| > \frac{1}{2}$, $n = 1, 2, \ldots$. Let

$$M = \overline{[\{x_1, x_2, \ldots\}]}.$$

We claim that $M = X$. Suppose $M \neq X$. Let $x_0 \in X - M$. Then

$$d = d(x_0, M) > 0.$$

In view of Theorem 25, there exists a bounded linear functional $F$ such that $\|F\| = 1$, $F(x_0) \neq 0$ and if $x \in M$, $F(x) = 0$. Since $\|F\| = 1$, $F \in S$ and $F(x_n) = 0$, $n = 1, 2, \dots$, since $x_n \in M$, $n = 1, 2, \dots$. Now,

$$\begin{aligned} |f_n(x_n)| &= |\{f_n(x_n) - F(x_n)\} + F(x_n)| \\ &\leq |f_n(x_n) - F(x_n)| + |F(x_n)| \\ &= |(f_n - F)(x_n)|, \text{ since } F(x_n) = 0 \\ &\leq \|f_n - F\| \|x_n\|. \end{aligned}$$

Thus,

$$\begin{aligned} \frac{1}{2} < |f_n(x_n)| &\leq \|f_n - F\| \|x_n\| \\ &= \|f_n - F\|, \text{ since } \|x_n\| = 1, \end{aligned}$$

$$\textit{i.e.,}\ \frac{1}{2} < \|f_n - F\|, n = 1, 2, \dots,$$

which contradicts the fact that $\{f_1, f_2, \dots\}$ is a dense subset of $S$. So $M = X$, proving our claim. We now appeal to Theorem 26 to complete the proof. □

---

## 3.4 Non-uniqueness and uniqueness of the Hahn–Banach extension

We note that the Hahn–Banach extension need not be unique. We now give examples of non-unique Banach extensions.

**Example 24.** *Let $X = \mathbb{R}^2$. For $x = (x_1, x_2) \in \mathbb{R}^2$, define*

$$\|x\| = |x_1| + |x_2|.$$

*Let*

$$Y = \{(x_1, x_2) \in \mathbb{R}^2 / x_2 = 0\}$$

*and*

$$g((y_1, y_2)) = y_1, \ \text{for } y = (y_1, y_2) \in Y.$$

*It is easy to check that* $Y$ *is a subspace of* $X$ *and* $g$ *is a linear functional on* $Y$. *Now,*

$$\begin{aligned} |g(y)| = |g((y_1, y_2))| &= |y_1| \\ &\leq |y_1| + |y_2| \\ &= \|y\|, \end{aligned}$$

*which shows that* $g$ *is bounded. Define* $f_1, f_2$ *on* $X$ *by*

$$f_1((x_1, x_2)) = x_1 + x_2$$

*and*

$$f_2((x_1, x_2)) = x_1 - x_2, \ \text{where } x = (x_1, x_2) \in X.$$

*We can verify that both* $f_1, f_2$ *are norm-preserving extensions of* $g$.

**Example 25.** *Let* $X = \mathbb{R}^2$. *For* $x = (x_1, x_2) \in \mathbb{R}^2$, *define*

$$\|x\| = \max(\|x_1\|, \|x_2\|).$$

*Let*

$$Y = \{(x_1, x_2) \in \mathbb{R}^2 / x_1 = x_2\}.$$

*Define the functional* $g$ *on* $Y$ *by*

$$g((y_1, y_2)) = y_1, \ y = (y_1, y_2) \in Y.$$

*Now, define the functionals* $f_1, f_2$ *on* $X$ *by*

$$f_1((x_1, x_2)) = x_1$$

*and*

$$f_2((x_1, x_2)) = x_2.$$

*It is left to the reader to verify that $Y$ is a subspace of $X$, $g$ is a bounded linear functional on $Y$, and $f_1, f_2$ are both norm-preserving extensions of $g$.*

The following example shows that there can exist infinitely many Hahn–Banach extensions.

**Example 26.** *Let $X = C[0,1]$ with*

$$\|f\| = \sup_{t\in[0,1]} |f(t)|, f \in X.$$

*Let $Y$ be the subspace of $X$ of all constant functions.*
*Define the functional $g$ on $Y$ by*

$$g(y) = y(0), y \in Y.$$

*For any fixed $t \in [0,1]$, define*

$$f_t(x) = x(t), x \in X.$$

*The reader can verify that $g$ is a bounded linear functional on $Y$ and for every $t \in [0,1]$, $f_t$ is a norm-preserving extension of $g$ on the space $X$.*

So we naturally ask the question: When is the norm-preserving extension unique?
Theorem 28, proved below, answers this question.
The following definition is needed.

**Definition 39.** *A normed linear space $X$ is said to be "strictly convex," if $\|\frac{x+y}{2}\| < 1$, whenever $x, y \in X$, $x \neq y$, $\|x\| = \|y\| = 1$.*

Interpreting geometrically, $X$ is strictly convex if the midpoint of the line segment joining any two points on the unit sphere in $X$ does not lie on the unit sphere.

$\mathbb{R}^n$ or $\mathbb{C}^n$, with $\|.\|_p$, is strictly convex if $1 < p < \infty$ and not strictly convex if $p = 1$ or $p = \infty$.

We now characterize normed linear spaces that admit unique Hahn–Banach extensions.

**Theorem 28** (Taylor–Foguel). *Let $X$ be a normed linear space. Then every bounded linear functional on every subspace of $X$ has a unique norm-preserving linear extension to $X$ if and only if $\tilde{X}$ is strictly convex.*

*Proof.* Let $\tilde{X}$ be strictly convex. Let $Y$ be a subspace of $X$ and $g \in \tilde{Y}$. Let $f_1, f_2$ be norm-preserving extensions of $g$ to $X$. Then $\frac{f_1+f_2}{2}$ is a bounded linear functional on $X$, extending $g$. Hence,

$$\begin{aligned}\|g\| &\leq \left\|\frac{f_1 + f_2}{2}\right\| \\ &\leq \frac{1}{2}(\|f_1\| + \|f_2\|) \\ &= \frac{1}{2}(\|g\| + \|g\|) \\ &= \|g\|,\end{aligned}$$

from which we have

$$\left\|\frac{f_1 + f_2}{2}\right\| = \|g\|. \tag{3.12}$$

Since $\tilde{X}$ is strictly convex, we have $f_1 = f_2$ because, if $f_1 \neq f_2$, $f_1, f_2 \in \tilde{X}$, $\|f_1\| = \|f_2\| = 1$, then,

$$\left\|\frac{f_1 + f_2}{2}\right\| < 1 = \|g\|,$$

a contradiction of (3.12). Thus $X$ admits unique Hahn–Banach extensions.

Conversely, let $X$ admit unique Hahn–Banach extensions. We claim that $\tilde{X}$ is strictly convex. Suppose not. Then, there exist $f_1, f_2 \in \tilde{X}$, $f_1 \neq f_2$, $\|f_1\| = \|f_2\| = \left\|\frac{f_1+f_2}{2}\right\| = 1$.

Let

$$Y = \{x \in X / f_1(x) = f_2(x)\}.$$

Then $Y$ is a subspace of $X$. Since $f_1 \neq f_2$, there exists $a \in X$ such that $f_1(a) = 1$ and $f_2(a) \neq 1$. Since $\left\|\frac{f_1+f_2}{2}\right\| = 1$, there exists a sequence $\{x_n\}$ in $X$ such that $\|x_n\| = 1$, $n = 1, 2, \ldots$ and

$$|f_1(x_n) + f_2(x_n)| \to 2, n \to \infty.$$

Again, since $|f_1(x_n)|$, $|f_2(x_n)| \leq 1$, for $n = 1, 2, \ldots$, we can assume, if necessary, by passing on to a subsequence of $\{x_n\}$ and multiplying $x_n$ by a scalar of absolute value 1 that

$$f_1(x_n) \to 1, n \to \infty$$

and

$$f_2(x_n) \to 1, n \to \infty.$$

Let

$$k_n = \frac{f_1(x_n) - f_2(x_n)}{1 - f_2(a)}, n = 1, 2, \ldots.$$

Then, $k_n \to 0$, $n \to \infty$, and there exists a positive integer $n_0$ such that

$$\|x_n - k_n a\| \neq 0, \text{ for all } n \geq n_0.$$

For $n \geq n_0$, let

$$y_n = \frac{x_n - k_n a}{\|x_n - k_n a\|}.$$

Then

$$\begin{aligned}
f_1(y_n) &= \frac{1}{\|x_n - k_n a\|}[f_1(x_n) - k_n f_1(a)] \\
&= \frac{1}{\|x_n - k_n a\|}[f_1(x_n) - k_n], \text{ since } f_1(a) = 1 \\
&= \frac{1}{\|x_n - k_n a\|}\left[f_1(x_n) - \frac{f_1(x_n) - f_2(x_n)}{1 - f_2(a)}\right] \\
&= \frac{1}{\|x_n - k_n a\|}\left[\frac{f_1(x_n) - f_1(x_n)f_2(a) - f_1(x_n) + f_2(x_n)}{1 - f_2(a)}\right] \\
&= \frac{1}{\|x_n - k_n a\|}\left[\frac{f_2(x_n) - f_1(x_n)f_2(a)}{1 - f_2(a)}\right];
\end{aligned}$$

and

$$\begin{aligned} f_2(y_n) &= \frac{1}{\|x_n - k_n a\|}[f_2(x_n) - k_n f_2(a)] \\ &= \frac{1}{\|x_n - k_n a\|}\left[f_2(x_n) - \frac{f_1(x_n) - f_2(x_n)}{1 - f_2(a)} f_2(a)\right] \\ &= \frac{1}{\|x_n - k_n a\|}\left[\frac{f_2(x_n) - f_2(x_n)f_2(a) - f_1(x_n)f_2(a) + f_2(x_n)f_2(a)}{1 - f_2(a)}\right] \\ &= \frac{1}{\|x_n - k_n a\|}\left[\frac{f_2(x_n) - f_1(x_n)f_2(a)}{1 - f_2(a)}\right], \end{aligned}$$

so that

$$f_1(y_n) = f_2(y_n).$$

Thus, $y_n \in Y$, for all $n \geq n_0$. Already, $\|y_n\| = 1$, for all $n \geq n_0$. Since $k_n \to 0$, $f_1(x_n) \to 1$, $f_2(x_n) \to 1$, $n \to \infty$, we see that

$$f_1(y_n) \to 1, \; f_2(y_n) \to 1, \; n \to \infty.$$

Hence,

$$\|f_1/Y\| = \|f_2/Y\| = 1,$$

where $f_1/Y, f_2/Y$ denote the restrictions of $f_1, f_2$ on $Y$, respectively. This contradicts the uniqueness of the Hahn–Banach extensions on $X$. This proves that $\tilde{X}$ is strictly convex, completing the proof of the theorem. □

## Exercises

1. If $X$ is not the trivial space, prove that the conjugate space is not trivial, i.e., prove that non-zero, bounded, linear functionals exist on any non-trivial normed linear space.

2. If all the bounded linear functionals on a normed linear space vanish on a given vector, prove that the vector must be zero.

3. Let $X$ be a normed linear space. Let $M$ be a closed subspace of $X$. Let $x_0 \in X - M$. Prove that there exists a bounded linear functional $f$ on $X$ such that $f(x_0) \neq 0$ and $f(y) = 0$, for all $y \in M$.

4. Prove that the converse of Theorem 27 is false.

5. Give an example of a separable normed linear space and a non-separable one, justifying your claims.

6. Prove that $\mathbb{R}^n$ with norm $\|.\|_p$ is strictly convex if $1 < p < \infty$.

7. Prove that $\mathbb{R}^n$ with $\|.\|_p$ is not strictly convex if $p = 1$ or $p = \infty$.

8. Let $X$ be a normed linear space. If $f(x) = f(y)$, for all $f \in \tilde{X}$, prove that $x = y$.

9. Let $X$ be a normed linear space and $x \in X$. Prove that

$$\|x\| = \sup_{\substack{f \in \tilde{X} \\ \|f\|=1}} |f(x)|.$$

10. Let $X$ be a normed linear space. If $\tilde{X}$ is strictly convex, prove that for every $x \in X$, $x \neq 0$, there exists a unique $f \in \tilde{X}$ such that

$$f(x) = \|x\| \text{ and } \|f\| = 1.$$

# Chapter 4

# Reflexivity

## 4.1 Second conjugate spaces, reflexivity

We know that the set of all bounded linear functionals on a normed linear space $X$ is itself a normed linear space $\tilde{X}$. So we can talk of the second conjugate space $\tilde{\tilde{X}} = \widetilde{(\tilde{X})}$, the set of all bounded linear functionals on $\tilde{X}$, the third conjugate space $\tilde{\tilde{\tilde{X}}}$, and so on.

Let $y$ be a linear functional on the linear space $X$. We shall now denote by $[x, y]$, the element $y(x)$, where $x \in X$. We note that $[\ .\ ]$ is bilinear, i.e., it is linear in both the arguments,
i.e.,

$$[\alpha_1 x_1 + \alpha_2 x_2, y] = \alpha_1[x_1, y] + \alpha_2[x_2, y]$$

and

$$[x, \alpha_1 y_1 + \alpha_2 y_2] = \alpha_1[x, y_1] + \alpha_2[x, y_2].$$

Keeping $x$ fixed and allowing $y$ to vary over the linear space of all linear functionals on $X$, the bilinearity of $[\ .\ ]$ gives us a linear functional on the space of all linear functionals on $X$. Further, if $X$ is a normed linear space and $y$ varies over $\tilde{X}$, the above process yields a bounded linear functional on $\tilde{X}$. Consider the mapping

$$J : X \to \tilde{\tilde{X}}$$

$$x \to x'',$$

where the linear functional $x''$ is defined as follows:

Let $x' \in \tilde{X}$. Then

$$x''(x') = [x', x''] = [x, x'].$$

$x''$ is a linear functional on $\tilde{X}$. $x''$ is bounded, because

$$\|x''(x')\| = \|[x, x']\| = |x'(x)| \leq \|x'\|\|x\|,$$

from which we have

$$\|x''\| \leq \|x\|. \tag{4.1}$$

We claim that $\|x''\| = \|x\|$.

Case 1. If $x = 0$, $0 \leq \|x''\| \leq 0$, which implies that

$$\|x''\| = 0 = \|x\|.$$

Case 2. If $x \neq 0$, in view of Theorem 24, there exists a bounded linear functional $x_0'$ such that

$$\|x_0'\| = 1 \text{ and } x_0'(x) = \|x\|.$$

But

$$\begin{aligned} \|x''\| &= \sup_{\|x'\|=1} \|x''(x')\| \\ &= \sup_{\|x'\|=1} |x'(x)|. \end{aligned}$$

Now,

$$\begin{aligned} \|x\| &= x_0'(x) \\ &\leq \sup_{\|x'\|=1} |x'(x)| \\ &= \|x''\|. \end{aligned} \tag{4.2}$$

From (4.1) and (4.2), we have

$$\|x''\| = \|x\|,$$

$$i.e., \|x''\| = \|J(x)\| = \|x\|,$$

from which we conclude that $J$ is an isometry.

**Definition 40.** *If the mapping $J$ is surjective, i.e., onto, then $X$ is said to be "reflexive," i.e., $X$ is reflexive if $X$ is "congruent" to $\tilde{\tilde{X}}$, where "congruence" means "isometric isomorphism."*

---

## 4.2 Conjugate space of $\ell_p$, $1 < p < \infty$

**Definition 41.** *Let $\alpha \in \mathbb{C}$. Then signum $\alpha$, denoted by sgn $\alpha$, is defined by*

$$sgn\ \alpha = \begin{cases} 0, & if\ \alpha = 0; \\ \frac{\alpha}{|\alpha|}, & if\ \alpha \neq 0. \end{cases}$$

*It is immediate that*

$$|sgn\ \alpha| = \begin{cases} 0, & if\ \alpha = 0; \\ 1, & if\ \alpha \neq 0, \end{cases}$$

*and if $\alpha \neq 0$,*

$$\alpha\ sgn\ \overline{\alpha} = \alpha \frac{\overline{\alpha}}{|\alpha|} = \frac{|\alpha|^2}{|\alpha|} = |\alpha|,$$

*so that*

$$\alpha\ sgn\ \overline{\alpha} = \begin{cases} 0, & if\ \alpha = 0; \\ |\alpha|, & if\ \alpha \neq 0. \end{cases}$$

**Theorem 29.** *If $1 < p < \infty$, $\tilde{\ell}_p$ is congruent, i.e., isometrically isomorphic to $\ell_q$, where $\frac{1}{p} + \frac{1}{q} = 1$.*

*Proof.* Let $x = \{\xi_n\} \in \ell_p$. Then

$$\sum_{n=1}^{\infty} |\xi_n|^p < \infty.$$

Let, for $n = 1, 2, \ldots$, $e_n = \{0, 0, \ldots, 0, 1, 0, \ldots\}$, 1 occurring in the $n$th place and 0, elsewhere. We then have $e_n \in \ell_p$, $\|e_n\|_p = 1$, $n = 1, 2, \ldots$. Now,

$$x - \sum_{k=1}^{n} \xi_k e_k = \{0, 0, \ldots, 0, \xi_{n+1}, \xi_{n+2}, \ldots\},$$

so that

$$\left\| x - \sum_{k=1}^{n} \xi_k e_k \right\|_p = \left( \sum_{k=n+1}^{\infty} |\xi_k|^p \right)^{\frac{1}{p}}$$
$$\to 0,\ n \to \infty.$$

Thus,

$$x = \sum_{k=1}^{\infty} \xi_k e_k.$$

Let $f \in \tilde{\ell}_p$. Then

$$f\left( \sum_{k=1}^{n} \xi_k e_k \right) = \sum_{k=1}^{n} \xi_k f(e_k),$$

from which it follows that

$$f(x) = \sum_{k=1}^{\infty} \xi_k f(e_k).$$

Let $f \in \tilde{\ell}_p$ and $\alpha_k = f(e_k)$, $k = 1, 2, \ldots$. Consider the mapping

$$T : \tilde{\ell}_p \to \ell_q$$
$$f \to \{\alpha_n\}.$$

We first show that $T$ maps $\tilde{\ell}_p$ into $\ell_q$. Consider

$$x = \{\beta_k\} \in \ell_p,$$

where

$$\beta_k = \begin{cases} |\alpha_k|^{q-1}\ sgn\ \overline{\alpha}_k, & \text{if } 1 \le k \le n; \\ 0, & \text{if } k > n, \end{cases}$$

$\alpha_k$ as mentioned above. For $1 \le k \le n$,

$$\begin{aligned}
|\beta_k| &= |\alpha_k|^{q-1}, \\
i.e., |\beta_k|^p &= |\alpha_k|^{(q-1)p} = |\alpha|_k^q, \quad \text{since } \frac{1}{p} + \frac{1}{q} = 1, \\
i.e., \alpha_k \beta_k &= |\alpha_k|^{q-1} \alpha_k \ sgn\ \overline{\alpha}_k \\
&= |\alpha_k|^{q-1} |\alpha_k| \\
&= |\alpha_k|^q \\
&= |\beta_k|^p.
\end{aligned}$$

Now,

$$\|x\|_p = \left( \sum_{k=1}^{n} |\beta_k|^p \right)^{\frac{1}{p}} = \left( \sum_{k=1}^{n} |\alpha_k|^q \right)^{\frac{1}{p}}.$$

Since $x = \sum_{k=1}^{n} \beta_k e_k$,

$$\begin{aligned}
f(x) &= \sum_{k=1}^{n} \beta_k f(e_k) \\
&= \sum_{k=1}^{n} \alpha_k \beta_k, \quad \text{since } \alpha_k = f(e_k),\ k = 1, 2, \ldots, n, \\
&= \sum_{k=1}^{n} |\alpha_k|^q.
\end{aligned}$$

But for all $x \in \ell_p$,

$$|f(x)| \le \|f\| \|x\|_p.$$

Now,

$$\begin{aligned}
|f(x)| &= \sum_{k=1}^{n} |\alpha_k|^q \\
&\le \|f\| \left( \sum_{k=1}^{n} |\alpha_k|^q \right)^{\frac{1}{p}},
\end{aligned}$$

from which we have

$$\left(\sum_{k=1}^{n}|\alpha_k|^q\right)^{1-\frac{1}{p}} \leq \|f\|,$$

$$i.e., \left(\sum_{k=1}^{n}|\alpha_k|^q\right)^{\frac{1}{q}} \leq \|f\|,\ n = 1, 2, \ldots.$$

Consequently,

$$\left(\sum_{k=1}^{\infty}|\alpha_k|^q\right)^{\frac{1}{q}} \leq \|f\|,$$

which implies that $\{\alpha_k\} \in \ell_q$ and

$$\|T(f)\|_q \leq \|f\|. \tag{4.3}$$

For $x = \{\xi_k\} \in \ell_p$,

$$x = \sum_{k=1}^{\infty}\xi_k e_k,$$

$$f(x) = \sum_{k=1}^{\infty}\xi_k f(e_k)$$

$$= \sum_{k=1}^{\infty}\xi_k \alpha_k,$$

so that

$$|f(x)| \leq \sum_{k=1}^{\infty}|\xi_k||\alpha_k|$$

$$\leq \left(\sum_{k=1}^{\infty}|\alpha_k|^q\right)^{\frac{1}{q}}\left(\sum_{k=1}^{\infty}|\xi_k|^p\right)^{\frac{1}{p}}, \text{ by Hölder's inequality}$$

$$= \left(\sum_{k=1}^{\infty}|\alpha_k|^q\right)^{\frac{1}{q}}\|x\|_p.$$

Thus,

$$\|f\| \leq \left(\sum_{k=1}^{\infty}|\alpha_k|^q\right)^{\frac{1}{q}} = \|T(f)\|_q. \tag{4.4}$$

From (4.3) and (4.4), we get

$$\|T(f)\|_q = \|f\|,$$

so that $T$ is an isometry. It is clear that $T$ is linear and one-to-one, i.e., injective. We shall show that $T$ is onto, i.e., surjective. Let $\{\beta_k\} \in \ell_q$ and $x = \{\xi_k\} \in \ell_p$. Let

$$g(x) = \sum_{k=1}^{\infty} \xi_k \beta_k.$$

It is clear that $g$ is well defined and $g$ is linear. We claim that $g$ is bounded. Now,

$$\begin{aligned} |g(x)| &\leq \sum_{k=1}^{\infty} |\xi_k||\beta_k| \\ &\leq \left(\sum_{k=1}^{\infty} |\xi_k|^p\right)^{\frac{1}{p}} \left(\sum_{k=1}^{\infty} |\beta_k|^q\right)^{\frac{1}{q}}, \quad \text{by Hölder's inequality} \\ &= \|x\|_p \left(\sum_{k=1}^{\infty} |\beta_k|^q\right)^{\frac{1}{q}}, \end{aligned}$$

which shows that $g$ is bounded. Note that $g(e_k) = \beta_k$ and $T(g) = \{\beta_k\}$, proving that $T$ is onto. Thus,

$$\tilde{\ell}_p = \ell_q \text{ up to isometric isomorphism,}$$

completing the proof. □

In the context of Theorem 29, we note the following points:

**Note 1.** *For $x = \{\xi_k\} \in \ell_p$ and $f \in \tilde{\ell}_p$, we have the unique representation of $f$ as*

$$f(x) = \sum_{k=1}^{\infty} \xi_k f(e_k), \quad \textit{where } x = \sum_{k=1}^{\infty} \xi_k e_k.$$

**Note 2.**

$$\|f\| = \left(\sum_{k=1}^{\infty} |f(e_k)|^q\right)^{\frac{1}{q}}.$$

**Note 3.** *The Hilbert space $\ell_2$ is reflexive. More generally, we can prove that if $X$ is a Hilbert space, then $X$ is reflexive.*

## 4.3 Conjugate spaces of $\ell_1, c_0$

**Theorem 30.** *$\tilde{\ell}_1 = \ell_\infty$, where, as usual, $\ell_1$ is the Banach space of all sequences $x = \{x_n\}$ such that*

$$\sum_{n=1}^{\infty} |x_n| < \infty$$

*with $\|x\| = \sum_{n=1}^{\infty} |x_n|$ and $\ell_\infty$ is the Banach space of all bounded sequences $x = \{x_n\}$ with $\|x\| = \sup_n |x_n|$. We use the symbol $=$ in Theorem 30 in the congruence sense.*

*Proof.* Let, as usual,

$$e_n = \{0, 0, \ldots, 0, 1, 0, 0, \ldots\},$$

1 occurring in the $n$th place and 0, elsewhere. Clearly $e_n \in \ell_1$, $\|e_n\| = 1$, $n = 1, 2, \ldots$. Let $x = \{x_n\} \in \ell_1$ and $s_n = \sum_{k=1}^{n} \xi_k e_k$, $n = 1, 2, \ldots$. Now,

$$\|x - s_n\| = \sum_{k=n+1}^{\infty} |\xi_k|.$$

Since $\sum_{k=1}^{\infty} |\xi_k| < \infty$, given $\epsilon > 0$, there exists a positive integer $N$ such that

$$\|x - s_n\| < \epsilon, \quad \text{for all} \quad n \geq N.$$

Let $f \in \tilde{\ell}_1$ and $\mu_n = f(e_n)$, $n = 1, 2, \ldots$. Then,

$$|\mu_n| = |f(e_n)| \leq \|f\|\|e_n\| = \|f\|, \text{ since } \|e_n\| = 1, \ n = 1, 2, \ldots,$$

which implies that $\{\mu_n\} \in \ell_\infty$. Consider the mapping

$$T : \tilde{\ell}_1 \to \ell_\infty$$

$$f \to \{\mu_n\}, \quad \text{where } \mu_n = f(e_n), n = 1, 2, \ldots.$$

Clearly $T$ is linear. Also,

$$\|T(f)\| = \sup_n |\mu_n| \leq \|f\|. \tag{4.5}$$

Also,

$$\begin{aligned} |f(x - s_n)| &\leq \|f\|\|x - s_n\| \\ &\leq \|f\|\epsilon, \quad \text{for all } n \geq N. \end{aligned}$$

However,

$$\begin{aligned} f(x - s_n) &= f(x) - f(s_n) \\ &= f(x) - f\left(\sum_{k=1}^{n} \xi_k e_k\right) \\ &= f(x) - \sum_{k=1}^{n} \xi_k \mu_k, \end{aligned}$$

so that

$$\left| f(x) - \sum_{k=1}^{n} \xi_k \mu_k \right| < \epsilon \|f\|, \text{ for all } n \geq N,$$

from which it follows that

$$f(x) = \sum_{k=1}^{\infty} \xi_k \mu_k.$$

It is also clear that $f = 0$ implies that $T(f) = 0$. Hence the mapping $T$ is one-to-one. Let $g = \{\mu_n\} \in \ell_\infty$. Since $\{\mu_n\}$ is bounded, $\sum_{k=1}^{\infty} \xi_k \mu_k$ is absolutely convergent for every $\{\xi_n\} \in \ell_1$. Consider the functional $f$ on $\ell_1$ defined by

$$f(x) = \sum_{k=1}^{\infty} \xi_k \mu_k, \text{ for every } x = \{\xi_n\} \in \ell_1.$$

Clearly $f$ is linear on $\ell_1$. Also, for all $x = \{\xi_n\} \in \ell_1$,

$$\begin{aligned} |f(x)| &= \left|\sum_{k=1}^{\infty} \xi_k \mu_k\right| \\ &\leq \sum_{k=1}^{\infty} |\xi_k||\mu_k| \\ &\leq \|g\| \sum_{k=1}^{\infty} |\xi_k| \\ &= \|g\|\|x\|. \end{aligned}$$

Thus, $f$ is bounded, i.e., $f \in \tilde{\ell}_1$ and so $f$ is onto. Also,

$$\|f\| \leq \|g\|.$$

Since $T$ is onto, let $f \in \tilde{\ell}_1$ such that $g = T(f)$ so that

$$\|f\| \leq \|T(f)\|. \tag{4.6}$$

From (4.5) and (4.6), we have

$$\|T(f)\| = \|f\|$$

and so $T$ is an isometry. Consequently, $T$ is a congruence, completing the proof. □

**Theorem 31.** $\tilde{c}_0 = \ell_1$, *where* $c_0$ *is the Banach space of all null sequences under the norm*

$$\|x\| = \sup_n |x_n|, \quad x = \{x_n\} \in c_0$$

*(note that* $c_0$ *is a closed subspace of* $\ell_\infty$*).*

*Proof.* As usual, let

$$e_n = \{0, 0, \ldots, 0, 1, 0, \ldots\},$$

1 occurring in the $n$th place and 0, elsewhere, $n = 1, 2, \ldots$. Clearly $e_n \in c_0$, $\|e_n\| = 1$, $n = 1, 2, \ldots$. Let $x = \{\xi_n\} \in c_0$ and $s_n = \sum_{k=1}^{n} \xi_k e_k$, $n = 1, 2, \ldots$. Since $\lim_{n\to\infty} \xi_n = 0$, given $\epsilon > 0$, there exists a positive integer $N$ such that

$$|\xi_n| < \epsilon, \quad \text{for all } n \geq N.$$

Then

$$\begin{aligned}\|x - s_n\| &= \sup_{k \geq n+1} |\xi_k| \\ &\leq \epsilon, \quad \text{for all } n \geq N.\end{aligned}$$

Now, let $f \in \tilde{c}_0$ and $\mu_n = f(e_n)$, $n = 1, 2, \ldots$.

$$\begin{aligned}|f(x - s_n)| &\leq \|f\|\|x - s_n\| \\ &\leq \|f\|\epsilon, \quad \text{for all } n \geq N,\end{aligned}$$

$$i.e., \left| f(x) - \sum_{k=1}^{n} \xi_k \mu_k \right| \leq \|f\|\epsilon, \quad \text{for all } n \geq N,$$

from which it follows that

$$f(x) = \sum_{k=1}^{\infty} \xi_k \mu_k.$$

We claim that $\{\mu_n\} \in \ell_1$. To prove the claim, let

$$\alpha_k = \begin{cases} 0, & \text{if } \mu_k = 0; \\ \frac{|\mu_k|}{\mu_k}, & \text{if } \mu_k \neq 0, \end{cases} \qquad k = 1, 2, \ldots.$$

Let $x_n = \sum_{k=1}^{n} \alpha_k e_k$, $n = 1, 2, \ldots$. Clearly $\|x_n\| \leq 1$, $n = 1, 2, \ldots$ and $f(x_n) = \sum_{k=1}^{n} |\mu_k|$, $n = 1, 2, \ldots$. Now,

$$\sum_{k=1}^{n} |\mu_k| = |f(x_n)| \leq \|f\|\|x_n\| \leq \|f\|,$$

from which it follows that $\sum_{k=1}^{\infty} |\mu_k| < \infty$ and so $\{\mu_n\} \in \ell_1$.

Also,

$$\sum_{k=1}^{\infty} |\mu_k| \leq \|f\|. \tag{4.7}$$

Let

$$\begin{aligned}T : \tilde{c}_0 &\to \ell_1 \\ f &\to \{\mu_n\},\end{aligned}$$

where $\mu_n = f(e_n)$, $n = 1, 2, \ldots$. The reader can verify that $T$ is linear and one-to-one. Let $\{\mu_n\} \in \ell_1$. Then $\sum_{k=1}^{\infty} \xi_k \mu_k$ converges absolutely, for every $x = \{\xi_n\} \in c_0$. Let $f$ be a functional on $c_0$ defined by

$$f(x) = \sum_{k=1}^{\infty} \xi_k \mu_k, \quad \text{for } x = \{\xi_n\} \in c_0.$$

$f$ is linear on $c_0$. Also,

$$\begin{aligned} |f(x)| = \left| \sum_{k=1}^{\infty} \xi_k \mu_k \right| &\leq \sum_{k=1}^{\infty} |\xi_k||\mu_k| \\ &\leq \|x\| \sum_{k=1}^{\infty} |\mu_k| \\ &= \|x\| \|T(f)\|. \end{aligned}$$

Thus,

$$\|f\| \leq \|T(f)\|. \tag{4.8}$$

From (4.7) and (4.8), we have

$$\|T(f)\| = \|f\|,$$

$$i.e., T \text{ is an isometry.}$$

In other words, $\tilde{c}_0$ and $\ell_1$ are congruent, completing the proof. □

---

## 4.4 Conjugate space of $C[a, b]$

We need the following concepts.

**Definition 42.** *Let $a, b \in \mathbb{R}$ with $a < b$. A finite subset $P$ of $[a, b]$, which contains $a$ and $b$ is called a "partition" of $[a, b]$. If $P$ is a partition of $[a, b]$ and when we write $P = \{t_0, t_1, \ldots, t_n\}$, it is understood that*

$$a = t_0 < t_1 < \cdots < t_{n-1} < t_n = b.$$

*Let $f, \alpha$ be real-valued, bounded functions on $[a, b]$. We say that $f$ is $\alpha$-integrable, if there exists a real number $I$ with the following property: for each $\epsilon > 0$, there exists $\delta > 0$ such that*

$$\left| I - \sum_{k=1}^{n} f(s_k)\{\alpha(t_k) - \alpha(t_{k-1})\} \right| < \epsilon,$$

*for all partitions $\{t_0, t_1, \ldots, t_n\}$ of $[a, b]$ with $t_k - t_{k-1} < \delta$, for $k = 1, 2, \ldots, n$ and for all real numbers $s_1, s_2, \ldots, s_n$ with $t_{k-1} \leq s_k \leq t_k$, for $k = 1, 2, \ldots, n$. If $f$ is $\alpha$-integrable, the number $I$ is unique and it is denoted by*

$$\int_a^b f d\alpha \quad or \quad \int_a^b f(t) d\alpha(t).$$

*It is called the Riemann–Stieltjes integral of $f$ with respect to $\alpha$.*

**Definition 43.** *Let $\alpha$ be a real-valued function defined on $[a, b]$. Given a partition $\{t_0, t_1, \ldots, t_n\}$ of $[a, b]$, let*

$$V(P, \alpha) = \sum_{k=1}^{n} |\alpha(t_k) - \alpha(t_{k-1})|.$$

*The function $\alpha$ is said to be of "bounded variation" on $[a, b]$ if the set*

$$\{V(P, \alpha)/P \text{ is a partition of } [a, b]\}$$

*is bounded. If $\alpha$ is of bounded variation on $[a, b]$, the number*

$$V(\alpha) = \sup_P V(P, \alpha), P \text{ being a partition of } [a, b]$$

*is called the "total variation" of $\alpha$ on $[a, b]$.*

The following result is easily proved.

**Theorem 32.** *If $\alpha$ is of bounded variation on $[a, b]$, then each $f \in C[a, b]$ is $\alpha$-integrable.*

We now take up the following important representation theorem.

**Theorem 33.** *Let $f \in \widetilde{C[a, b]}$. Then there exists a function $g \in BV[a, b]$ such that*

$$f(x) = \int_a^b x(t) dg(t), \quad \text{for all } x \in C[a, b]$$

*and such that*

$$\|f\| = V(g),$$

*where* $V(g)$ *denotes the total variation of* $g$ *on* $[a,b]$ *and* $BV[a,b]$ *is the class of all functions of bounded variation on* $[a,b]$.

*Proof.* $C[a,b]$ can be considered as a subspace of $B[a,b]$, the set of all bounded functions defined on $[a,b]$, which is a Banach space under the norm

$$\|f\| = \sup_{x\in[a,b]} |f(x)|, \ f \in B[a,b].$$

Let $f$ be a bounded linear functional on $C[a,b]$. By the Hahn–Banach theorem, there exists a bounded linear functional $F$ defined on $B[a,b]$, extending $f$, such that $\|F\| = \|f\|$. For $a < s \leq b$, consider the characteristic function of $[a,s]$, i.e., $\chi_{[a,s]}(t)$:

$$\chi_{[a,s]}(t) = \begin{cases} 1, & \text{if } t \in [a,s]; \\ 0, & \text{if } t \notin [a,s]. \end{cases}$$

It is clear that $\chi_{[a,s]} \in B[a,b]$, for every such $s$.
We now define

$$g(s) = F(\chi_{[a,s]}), \quad \text{for } a < s \leq b$$

and

$$g(a) = 0.$$

Since $F$ is defined on $B[a,b]$, the above definition is sensible. Let $\{t_0, t_1, \ldots, t_n\}$ be a partition of $[a,b]$. For $i = 1, 2, \ldots, n$, define

$$\alpha_i = \overline{sgn}\{g(t_i) - g(t_{i-1})\},$$

where the bar denotes the complex conjugate. We also define

$$y(t) = \begin{cases} \alpha_1, & \text{if } t_0 \leq t \leq t_1; \\ \alpha_i, & \text{if } t_{i-1} < t \leq t_i, i = 2, 3, \ldots, n. \end{cases}$$

It is clear that $|\alpha_i| = 0$ or $1$, $i = 1, 2, \ldots, n$ and so $y \in B[a, b]$. Also, $\|y\| \leq 1$. We now claim that

$$y(t) = \sum_{i=1}^{n} \alpha_i \{y_i(t) - y_{i-1}(t)\},$$

where

$$y_i = \chi_{[a,t_i]}, \; i = 1, 2, \ldots, n$$

and

$$y_0 = 0.$$

We prove the claim for $t_0 \leq t \leq t_1$, the proof of the other cases follow similarly. Now, for $t_0 \leq t \leq t_1$,

$$\begin{aligned}
&\sum_{i=1}^{n} \alpha_i \{y_i(t) - y_{i-1}(t)\} \\
&\qquad = \alpha_1(1-0) + \alpha_2(1-1) + \cdots + \alpha_n(1-1) \\
&\qquad = \alpha_1 \\
&\qquad = y(t).
\end{aligned}$$

Now,

$$\begin{aligned}
F(y) &= \sum_{i=1}^{n} \alpha_i [F(y_i) - F(y_{i-1})] \\
&= \sum_{i=1}^{n} \alpha_i [g(t_i) - g(t_{i-1})],
\end{aligned}$$

since

$$F(y_i) = F(\chi_{[a,t_i]}) = g(t_i).$$

Since $\alpha \overline{sgn\ \alpha} = |\alpha|$, using the definition of $\alpha_i$,

$$F(y) = \sum_{i=1}^{n} |g(t_i) - g(t_{i-1})|,$$

so that

$$\begin{aligned}\sum_{i=1}^{n}|g(t_i)-g(t_{i-1})| &= |F(y)|\\ &\leq \|F\|\|y\|\\ &\leq \|F\|, \text{ since } \|y\|\leq 1\\ &= \|f\|.\end{aligned}$$

Thus, $g$ is of bounded variation on $[a,b]$ and

$$V(g)\leq \|f\|. \tag{4.9}$$

Next, let $x\in C[a,b]$. Define

$$z(t)=\sum_{i=1}^{n}x(t_{i-1})\{y_i(t)-y_{i-1}(t)\}.$$

Clearly, $z\in B[a,b]$. Now,

$$F(z)=\sum_{i=1}^{n}x(t_{i-1})\{g(t_i)-g(t_{i-1})\}.$$

$$z(t)-x(t)=\begin{cases}x(t_0)-x(t), & \text{if } t_0\leq t\leq t_1;\\ x(t_{i-1})-x(t), & \text{if } t_{i-1}<t\leq t_i, i=2,3,\ldots,n.\end{cases}$$

Let $\eta=\max\limits_{1\leq i\leq n}|t_i-t_{i-1}|$. Since $x\in C[a,b]$, it is uniformly continuous on $[a,b]$ and hence

$$\|z-x\|\to 0,\ \eta\to 0.$$

Since $F$ is a bounded linear functional,

$$\lim_{\eta\to 0}F(z)=\lim_{z\to x}F(z)=F(x),$$

but

$$F(z)=\sum_{i=1}^{n}x(t_{i-1})\{g(t_i)-g(t_{i-1})\}.$$

Hence, as $\eta\to 0$,

$$F(x)=\int_a^b x(t)dg(t).$$

As $x \in C[a,b]$ is arbitrary, $F$ must agree with $f$ on $[a,b]$ and so

$$f(x) = \int_a^b x(t)dg(t), \quad \text{for all } x \in C[a,b]. \tag{4.10}$$

We now claim that

$$V(g) = \|f\|.$$

We know that

$$\begin{aligned} |f(x)| &\leq \max_{t \in [a,b]} |x(t)| V(g) \\ &= V(g)\|x\|, \end{aligned}$$

from which, we have

$$\|f\| \leq V(g). \tag{4.11}$$

Using (4.9) and (4.11), we get

$$\|f\| = V(g),$$

completing the proof of the theorem. □

**Remark 1.** *We are successful in proving a certain correspondence between the bounded linear functionals on $C[a,b]$ and the functions of bounded variation on $[a,b]$. We note that this correspondence is not well defined in the sense that there is no unique function of bounded variation on $[a,b]$ for which (4.10) holds for all $x \in C[a,b]$. We can add an arbitrary constant to $g$ without affecting (4.10) because the partial sums induced by $g+c$ are the same as those induced by $g$. In this context, we also note that if $h$ is a function of bounded variation on $[a,b]$ such that $h(a) = g(a)$, $h(b) = g(b)$ and $h(t) = g(t)$ at all interior points of $[a,b]$, where $g$ is continuous, we can show that*

$$\int_a^b x(t)dg(t) = \int_a^b x(t)dh(t),$$

*regardless of how $h$ is defined at the points of discontinuity of $g$ in $(a,b)$.*

Now our aim is to establish a congruence between $\widetilde{C[a,b]}$ and a suitable space of functions of bounded variation.

**Definition 44.** *For $x_1, x_2 \in BV[a,b]$, define*

$$x_1 \sim x_2$$

*if*

$$\int_a^b y(t)dx_1(t) = \int_a^b y(t)dx_2(t), \quad \textit{for all } y \in C[a,b].$$

We can easily prove that "$\sim$" is an equivalence relation.

**Theorem 34.** *$x \sim 0$ if and only if for any $c \in (a,b)$, $x(a) = x(b) = x(c+0) = x(c-0)$, $x \in BV[a,b]$.*

Necessity. Let $x \sim 0$. Let $y(t) = 1$, for all $t \in [a,b]$. Then

$$0 = \int_a^b dx(t) = x(b) - x(a),$$

$$i.e., x(a) = x(b).$$

It is easy to prove the following:

(i) for $a \leq c \leq b$, $h > 0$,

$$\frac{1}{h}\int_c^{c+h} x(t)dt \to x(c+0), \ h \to 0;$$

and

(ii) for $a < c \leq b$, $h > 0$,

$$\frac{1}{h}\int_{c-h}^{c} x(t)dt \to x(c-0), \ h \to 0.$$

Consider the continuous function

$$y(t) = \begin{cases} 1, & \text{if } a \leq t \leq c; \\ 1 - \frac{t-c}{h}, & \text{if } c < t \leq c+h; \\ 0, & \text{if } c+h < t \leq b. \end{cases}$$

Since $x \sim 0$,

$$\begin{aligned}
0 &= \int_a^b y(t)dx(t) \\
&= \int_a^c y(t)dx(t) + \int_c^b y(t)dx(t) \\
&= \int_a^c dx(t) + \int_c^{c+h} y(t)dx(t) \\
&= x(c) - x(a) + \int_c^{c+h} y(t)dx(t).
\end{aligned}$$

Using the formula of integration by parts,

$$\int_c^{c+h} y(t)dx(t) = -x(c) + \frac{1}{h}\int_c^{c+h} x(t)dt,$$

so that

$$\begin{aligned}
0 &= x(c) - x(a) - x(c) + \frac{1}{h}\int_c^{c+h} x(t)dt \\
&= -x(a) + \frac{1}{h}\int_c^{c+h} x(t)dt, \\
i.e., x(a) &= \frac{1}{h}\int_c^{c+h} x(t)dt.
\end{aligned}$$

Allowing $h \to 0$, we have

$$x(a) = x(c+0).$$

Similarly, we get

$$x(b) = x(c-0).$$

Sufficiency. For $c \in (a, b)$, let

$$x(a) = x(b) = x(c+0) = x(c-0), \quad \text{where } x \in BV[a, b].$$

Define $\hat{x}(t) = x(a)$, for all $t \in [a, b]$. Then

$$\hat{x}(b) = x(a) = x(b).$$

Also, $\hat{x}(t) = x(t)$ at all interior points of $[a, b]$ at which $x$ is continuous. Hence,

$$\int_a^b y(t)dx(t) = \int_a^b y(t)d\hat{x}(t), \quad \text{for all } y \in C[a, b].$$

Since $\hat{x}$ is a constant, total variation of $\hat{x}$ is 0. We know that if

$$f(y) = \int_a^b y(t)d\hat{x}(t),$$

then

$$\begin{aligned}\|f\| &= V(\hat{x}) \\ &= 0,\end{aligned}$$

which implies

$$f = 0,$$

$$i.e., f(y) = 0, \quad \text{for all } y \in C[a,b].$$

Thus,

$$\int_a^b y(t)dx(t) = 0, \quad \text{for all } y \in C[a,b].$$

Consequently,

$$x \sim 0,$$

completing the proof of the theorem. □

**Definition 45.** *The function $g \in BV[a,b]$ is said to be "normalized" if $g(a) = 0$ and $g(t+0) = g(t)$, for all $t \in [a,b]$.*

The set of all normalized functions of bounded variation on $[a,b]$ is denoted by $NBV[a,b]$ and note that $NBV[a,b]$ is a subspace of $BV[a,b]$.

**Theorem 35.** *If $x_1, x_2 \in NBV[a,b]$ and $x_1 \sim x_2$, then $x_1 = x_2$.*

*Proof.* $x_1 \sim x_2$ implies $x_1 - x_2 \sim 0$. Hence,

$$(x_1 - x_2)(a) = (x_1 - x_2)(b), \quad \text{by Theorem 34.}$$

Since $x_1, x_2$ are normalized, $x_1(a) = 0 = x_2(a)$, so that $x_1(b) = x_2(b)$. Again, since $x_1 - x_2 \sim 0$, for $c \in (a,b)$,

$$(x_1 - x_2)(a) = (x_1 - x_2)(c+0), \quad \text{again by Theorem 34,}$$

$$i.e., 0 = x_1(c+0) - x_2(c+0)$$

$$i.e., x_1(c) = x_2(c),$$

since $x_1, x_2$ are continuous from the right. Thus, $x_1(t) = x_2(t)$, for all $t \in [a, b]$. In other words, $x_1 = x_2$. □

**Theorem 36.** *Each equivalence class of $BV[a, b]$, under the equivalence relation $\sim$, contains exactly one normalized function.*

*Proof.* For a given $x \in BV[a, b]$, we will exhibit $\hat{x} \in NBV[a, b]$, such that

$$(x - \hat{x}) \sim 0.$$

$\hat{x}$ is unique in view of Theorem 35. Let $x \in BV[a, b]$. Define $\hat{x}$ as follows:

$$\hat{x}(a) = 0;$$
$$\hat{x}(t) = x(t + 0) - x(a), \; t \in (a, b);$$
$$\hat{x}(b) = x(b) - x(a).$$

The reader can easily prove that

$$\hat{x} \in NBV[a, b]$$

and

$$\hat{x} \sim x.$$

We now claim that

$$V(\hat{x}) \leq V(x).$$

To prove the claim, let $\{t_0, t_1, \ldots, t_n\}$ be a partition of $[a, b]$. Given $\epsilon > 0$, we can find $c_1, c_2, \ldots, c_{n-1}$, at which $x$ is continuous to the right of $t_1, t_2, \ldots, t_{n-1}$ such that

$$|x(t_i + 0) - x(c_i)| < \frac{\epsilon}{2n}, \; i = 1, 2, \ldots, (n - 1).$$

By taking $c_0 = a$ and $c_n = b$, we see that

$$\begin{aligned}\hat{x}(t_i) - \hat{x}(t_{i-1}) &= \{x(t_i + 0) - x(a)\} \\ &\quad - \{x(t_{i-1} + 0) - x(a)\} \\ &= x(t_i + 0) - x(t_{i-1} + 0) \\ &= \{x(t_i + 0) - x(c_i)\} \\ &\quad - \{x(t_{i-1} + 0) - x(c_{i-1})\} \\ &\quad + \{x(c_i) - x(c_{i-1})\},\end{aligned}$$

so that

$$\begin{aligned}|\hat{x}(t_i) - \hat{x}(t_{i-1})| &\leq \frac{\epsilon}{2n} + \frac{\epsilon}{2n} + |x(c_i) - x(c_{i-1})| \\ &= \frac{\epsilon}{n} + |x(c_i) - x(c_{i-1})|.\end{aligned}$$

Thus,

$$\begin{aligned}\sum_{i=1}^{n} |\hat{x}(t_i) - \hat{x}(t_{i-1})| &\leq \sum_{i=1}^{n} |x(c_i) - x(c_{i-1})| + \epsilon \\ &\leq V(x) + \epsilon\end{aligned}$$

and so

$$V(\hat{x}) \leq V(x) + \epsilon, \quad \text{for every } \epsilon > 0.$$

Since $\epsilon > 0$ is arbitrary,

$$V(\hat{x}) \leq V(x).$$

One can now see that $\hat{x}$ satisfies the requirements of the theorem, completing the proof. □

We now take up the main result of this section.

**Theorem 37.** *$\widetilde{C[a,b]}$ and $NBV[a,b]$ are congruent, where the congruence is given by*

$$\begin{aligned}T : NBV[a,b] &\to \widetilde{C[a,b]} \\ g(t) &\to f(x),\end{aligned}$$

*where*

$$f(x) = \int_a^b x(t)dg(t), \quad \textit{for all } x \in C[a,b].$$

*Proof.* Clearly $T$ is a linear transformation and $f$ is a linear functional. We also have

$$\left| \int_a^b x(t)dg(t) \right| \leq V(g)\|x\|,$$

$$i.e., |f(x)| \leq V(g)\|x\|,$$

from which we have

$$\|f\| \leq V(g).$$

Since

$$\|g\| = g(a) + V(g) \text{ and } g(a) = 0, \quad \text{for } g \in NBV[a,b],$$

$$V(g) = \|g\|,$$

so that

$$\|f\| \leq \|g\|. \tag{4.12}$$

$T$ is onto. Let $f \in \widetilde{C[a,b]}$. Then, by Theorem 33, there exists $h \in BV[a,b]$ such that

$$f(x) = \int_a^b x(t)dh(t), \quad \text{for all } x \in C[a,b].$$

Also, by Theorem 35 and Theorem 36, there exists a unique $g \in NBV[a,b]$ such that $h \sim g$. Thus,

$$f(x) = \int_a^b x(t)dg(t), \quad \text{for all } x \in C[a,b].$$

Clearly $T(g) = f$. Also, $V(g) \leq V(h) = \|f\|$,

$$i.e., V(g) \leq \|f\|. \tag{4.13}$$

From (4.12) and (4.13), we get

$$\|T(g)\| = \|f\| = V(g) = \|g\|.$$

$T$ is thus an isometry and hence one-to-one. Consequently, $T$ is a congruence, i.e.,

$$\widetilde{C[a,b]} = NBV[a,b], \quad \text{equality meaning congruence.}$$

Proof of the theorem is now complete. □

## Exercises

1. If $X$ is finite dimensional, prove that $X$ is reflexive.

2. If $X$ is reflexive, prove that $X$ is a Banach space.

3. Prove that $B[a,b]$ is a Banach space under the norm

$$\|f\| = \sup_{x \in [a,b]} |f(x)|, \; f \in B[a,b].$$

4. Prove that $BV[a,b]$ is a Banach space with respect to the norm

$$\|f\| = V(f) + |f(a)|, \; f \in BV[a,b].$$

We have used a few results on Riemann–Stieltjes integrals in the course of the above chapter. We leave these results as exercises (Exercises 5–10).

5. If $\alpha$ is of bounded variation on $[a,b]$, prove that each $f \in C[a,b]$ is $\alpha$-integrable.

6. If $\varphi$ is continuous on $[a,b]$ and $\psi$ is of bounded variation on $[a,b]$, prove that

$$\left| \int_a^b \varphi(t) d\psi(t) \right| \leq V(\psi) \max_{t \in [a,b]} |\varphi(t)|.$$

7. If $h$ is a function of bounded variation on $[a,b]$ such that $h(a) = g(a)$, $h(b) = g(b)$, and $h(t) = g(t)$ at all interior points of $[a,b]$ where $g$ is continuous, prove that

$$\int_a^b x(t) dg(t) = \int_a^b x(t) dh(t), \; x \in C[a,b]$$

regardless of how $h$ is defined at the points of discontinuity of $g$ in $(a,b)$.

8. If $c \in (a, b)$, prove that

$$\int_a^b f d\alpha = \int_a^c f d\alpha + \int_c^b f d\alpha.$$

9. $\int_a^b f(t) d\alpha(t)$ exists; prove that $\int_a^b \alpha(t) df(t)$ also exists and

$$\int_a^b f(t) d\alpha(t) + \int_a^b \alpha(t) df(t) = f(b)\alpha(b) - f(a)\alpha(a)$$

(formula of integration by parts).

10. If $f, f_1, f_2 \in C[a, b]$, $g, g_1, g_2 \in BV[a, b]$, prove that

(i) $\int_a^b (\alpha f_1 + \beta f_2) dg = \alpha \int_a^b f_1 dg + \beta \int_a^b f_2 dg$

and

(ii) $\int_a^b f d(\alpha g_1 + \beta g_2) = \alpha \int_a^b f dg_1 + \beta \int_a^b f dg_2$.

11. Prove that $\tilde{c} = \ell_1$ up to congruence.

12. Consider the complex Euclidean $n$-space $\mathbb{C}^n$. For $x = (x_1, x_2, \ldots, x_n) \in \mathbb{C}^n$, take $\|x\| = \sup_{1 \le i \le n} |\alpha_i|$. Find the conjugate space.

13. Give examples of spaces that are (i) reflexive and (ii) not reflexive, justifying your contention.

14. Let $X$ be a normed linear space. If $\tilde{X}$ is separable, prove that $X$ is separable. Is the converse true? Justify your answer.

15. Let $Y$ be a dense subspace of a normed linear space $X$. Prove that $\tilde{Y} = \tilde{X}$, up to congruence.

16. Prove that a closed subspace of a reflexive Banach space is reflexive.

17. If $X$ is a Banach space, prove that $X$ is reflexive if and only if $\tilde{X}$ is reflexive.

# Chapter 5

# Banach–Steinhaus Theorem

## 5.1 Baire's theorem: its consequences and applications

Throughout this section, $(X, d)$ denotes a metric space, $\mathcal{G}$ denotes the class of all open sets, and $\mathcal{F}$ denotes the class of all closed sets in $X$.

**Definition 46.** *$G_\delta$ denotes the class of all sets in $X$ that can be expressed as a countable intersection of open sets. Any member of this class is said to be of the type $G_\delta$.*

**Definition 47.** *$F_\sigma$ denotes the class of all sets in $X$ that can be expressed as a countable union of closed sets. Any member of this family is said to be of the type $F_\sigma$.*

**Note 4.** *A countable union of sets of typc $F_\sigma$ is again of type $F_\sigma$; a countable intersection of sets of type $G_\delta$ is again of type $G_\delta$.*

**Note 5.** *Any open set is of type $G_\delta$ and any closed set is of type $F_\sigma$, i.e., $\mathcal{G} \subset G_\delta$ and $\mathcal{F} \subset F_\sigma$.*

**Theorem 38.** *We have*

$$\mathcal{G} \subset F_\sigma \tag{5.1}$$

*and*

$$\mathcal{F} \subset G_\delta. \tag{5.2}$$

*Proof.* We first prove (5.2). Let $F \in \mathcal{F}$. For $n = 1, 2, \ldots$, let

$$O_n = \bigcup_{x \in F} S_{\frac{1}{n}}(x),$$

where $S_{\frac{1}{n}}(x)$ is the open sphere with $x$ as the center and $\frac{1}{n}$ as the radius. It is clear that $O_n$ is open and $O_n \supset F$, $n = 1, 2, \ldots$. Hence,

$$\bigcap_{n=1}^{\infty} O_n \supset F. \tag{5.3}$$

Now, let $y \in \bigcap_{n=1}^{\infty} O_n$. Then $y \in O_n = \bigcup_{x \in F} S_{\frac{1}{n}}(x)$, for every $n = 1, 2, \ldots$ and so $y \in S_{\frac{1}{n}}(x)$, for some $x \in F$. Now, given $\epsilon > 0$, there exists a positive integer $n$ such that $\frac{1}{n} < \epsilon$, so that there exists $x \in F$ such that $x \in S_\epsilon(y)$. Consequently,

$$y \in \overline{F} = F, \text{ since } F \text{ is closed.}$$

Thus,

$$\bigcap_{n=1}^{\infty} O_n \subset F. \tag{5.4}$$

From (5.3) and (5.4), we have

$$\bigcap_{n=1}^{\infty} O_n = F, \tag{5.5}$$

and so $F \in G_\delta$, i.e., $\mathcal{F} \subset G_\delta$. Taking complements in (5.5), (5.1) follows. □

**Definition 48.** *A set $A$ in a metric space $(X, d)$ is said to be "nowhere dense" if $(\overline{A})^\circ = \phi$, where $(\overline{A})^\circ$ denotes the interior of the set $\overline{A}$.*

**Definition 49.** *A subset $A$ of the metric space $(X, d)$ is said to be "somewhere dense," if it is not nowhere dense.*

We now make the following important observations.

**Note 6.** *Since $\overline{A}$ is closed, $A$ is nowhere dense in $X$ if and only if $\overline{A}$ is nowhere dense in $X$. A subset of a nowhere dense set is itself nowhere dense.*

**Definition 50.** *A set $A$ is said to be of the "first category" if it can be written as a countable union of nowhere dense sets. Otherwise, it is said to be of the "second category."*

**Note 7.** *A countable union of sets of the first category is again a set of the first category. The empty set is of the first category and consequently, any set of the second category is never empty. If $C = A \cup B$, where $A$ is of the first category and $C$ is of the second category, then $B$ must be of the second category.*

**Theorem 39.**

(i) *The set $\mathbb{Q}$ of rational numbers is of type $F_\sigma$ and is of the first category and*

(ii) *the set $\mathbb{Q}'$ of all irrational numbers is of type $G_\delta$ and is of the second category.*

*Proof.* Since $\mathbb{Q}$ is countable,

$$\mathbb{Q} = \bigcup_{i=1}^{\infty} \{x_i\}, \tag{5.6}$$

where $x_i$ is a rational number, $i = 1, 2, \ldots$. Since $\{x_i\}$ is closed, $\{x_i\}^\circ = \phi$. Hence $\mathbb{Q}$ is of type $F_\sigma$ and is of the first category. By taking complements in (5.6), one can see that $\mathbb{Q}'$, the set of all irrational numbers, is of type $G_\delta$. We know that

$$\mathbb{R} = \mathbb{Q} \cup \mathbb{Q}',$$

$\mathbb{R}$ being the set of all real numbers. We already know that $\mathbb{Q}$ is of the first category. Baire's theorem, proved below, implies that $\mathbb{R}$ is of the second category. By the observation made in Note 7, it follows that $\mathbb{Q}'$ is of the second category. □

We need the following result to prove Baire's theorem, which we state, leaving the proof to the reader.

By a nested sequence of sets $\{F_n\}$ we mean $F_1 \supset F_2 \supset \cdots \supset F_n \supset \ldots$.

**Theorem 40** (Nested sphere theorem). *A metric space $(X, d)$ is complete if and only if every nested sequence $\{S_n\} = \{S(x_n, r_n)\}$ of closed spheres in $X$ such that $r_n \to 0$, $n \to \infty$ has a non-empty intersection, i.e.,* $\bigcap_{n=1}^{\infty} S_n \neq \phi$.

We now take up Baire's theorem.

**Theorem 41** (Baire's theorem). *Any complete metric space is of the second category.*

*Proof.* We claim that no set of the first category can be equal to the entire space $X$. Let $A$ be any set in $X$ that is of the first category. We can now write

$$A = \bigcup_{n=1}^{\infty} E_n,$$

where $E_n$ is nowhere dense, $n = 1, 2, \ldots$.
Since $E_n$ is nowhere dense, it follows that

$$(\overline{E}_n)' = X, \; n = 1, 2, \ldots$$

(see Exercise 1 at the end of this chapter).
So there exist points in $(\overline{E}_n)'$. Let $x_1 \in (\overline{E}_1)'$. Since $(\overline{E}_1)'$ is open, there exists some neighborhood of $x_1$ lying wholly within $(\overline{E}_1)'$, i.e., there exits

$$S_\rho(x_1) \subset (\overline{E}_1)'.$$

For $\epsilon_1 < \rho$,

$$\overline{S_{\epsilon_1}(x_1)} \subset S_\rho(x_1) \subset (\overline{E}_1)' \subset E_1',$$

which implies that

$$\overline{S_{\epsilon_1}(x_1)} \cap E_1 = \phi.$$

We shall now use induction. Let $S_{\epsilon_k}(x_k)$, $k = 1, 2, \ldots, (n-1)$ be the spheres such that

$$\overline{S_{\epsilon_k}(x_k)} \cap E_k = \phi, \quad \text{where} \quad x_k \in (\overline{E}_k)'$$

and

$$\overline{S_{\epsilon_k}(x_k)} \subset \overline{S_{\epsilon_{k-1}}(x_{k-1})}, \; \epsilon_k \leq \frac{\epsilon_{k-1}}{2}.$$

We shall now construct the $n$th sphere having the above properties. We claim that

$$S_{\epsilon_{n-1}}(x_{n-1}) - \overline{E}_n \neq \phi.$$

If

$$S_{\epsilon_{n-1}}(x_{n-1}) - \overline{E}_n = \phi,$$

then

$$S_{\epsilon_{n-1}}(x_{n-1}) \subset \overline{E}_n$$

and so

$$x_{n-1} \in (\overline{E}_n)^\circ,$$

which is a contradiction, since $E_n$ is nowhere dense, proving the claim. Let

$$x_n \in S_{\epsilon_{n-1}}(x_{n-1}) - \overline{E}_n,$$

which implies that

$$x_n \in S_{\epsilon_{n-1}}(x_{n-1}) \cap (\overline{E}_n)' = B \text{ (say)}.$$

It is clear that $B$ is open. So, there exists an $\epsilon > 0$ such that

$$S_\epsilon(x_n) \subset B.$$

Let

$$\epsilon_n = \min\left(\epsilon, \frac{\epsilon_{n-1}}{2}\right).$$

Then

$$\overline{S_{\epsilon_n}(x_n)} \subset S_\epsilon(x_n) \subset B,$$

implying that

$$\overline{S_{\epsilon_n}(x_n)} \cap E_n = \phi.$$

Also we have

$$\overline{S_{\epsilon_n}(x_n)} \subset \overline{S_{\epsilon_{n-1}}(x_{n-1})} \text{ and } \epsilon_n \leq \frac{\epsilon_{n-1}}{2}.$$

Thus we have constructed the $n$th sphere. We now have a nested sequence of closed spheres in a complete metric space, whose diameters tend to 0. By

Theorem 40, there exists an $x_0$ such that $x_0 \in \bigcap_{n=1}^{\infty} \overline{S_{\epsilon_n}(x_n)}$. Since, $\overline{S_{\epsilon_n}(x_n)} \cap E_n = \phi$, $x_0 \notin E_n$, $n = 1, 2, \ldots$, i.e., $x_0 \in (E_n)'$, $n = 1, 2, \ldots$. But $A' = \bigcap_{n=1}^{\infty} E'_n$, so that $x_0 \in A'$, which, in turn, implies that $A \neq X$. So no set of the first category is equal to $X$. Consequently, $X$ is of the second category, completing the proof of the theorem. □

**Corollary 1.** *A complete metric space $X$, without isolated points, is uncountable. In particular, $\mathbb{R}$ is uncountable.*

We now take up for discussion some consequences and applications of Baire's theorem.

**Theorem 42.** *Let $(X, d)$ be a complete metric space and $A$ be an everywhere dense subset of $X$. If $A$ is of type $G_\delta$, then $A$ is of the second category.*

*Proof.* Since $A$ is of type $G_\delta$, there exist open sets $O_n$, $n = 1, 2, \ldots$ such that $A = \bigcap_{n=1}^{\infty} O_n$. Taking complements, we get

$$A' = \bigcup_{n=1}^{\infty} O'_n = \bigcup_{n=1}^{\infty} F_n, \quad \text{where } F_n = O'_n, n = 1, 2, \ldots.$$

It is clear that $A \subset O_n = F'_n$. Since

$$X = \overline{A} \subset \overline{(F'_n)} = (\overline{\overline{F}_n})' \subset X,$$

it follows that

$$X = (\overline{\overline{F}_n})'.$$

Thus, $F_n$ is nowhere dense, $n = 1, 2, \ldots$ and so $A'$ is of the first category. Since

$$X = A \cup A',$$

where $A'$ is of the first category and $X$ is of the second category, in view of Baire's theorem, it follows that $A$ is of the second category, using Note 7, completing the proof of the theorem. □

As a nice application of Baire's theorem, we shall prove the existence of functions defined on $[0, 1]$ that are continuous everywhere but differentiable nowhere.

**Theorem 43.** *There exist continuous functions on $[a, b]$ that are nowhere differentiable on $[a, b]$.*

*Proof.* Let $C_0[0, 1]$ be the metric space of all continuous functions defined on $[0, 1]$ for which $f(0) = f(1)$ with

$$d(f, g) = \sup_{x \in [0,1]} |f(x) - g(x)|, \ \ f, g \in C_0[0, 1]. \tag{5.7}$$

The reader can check that $C_0[0, 1]$ is complete and so it is of the second category, by Baire's theorem. We shall prove that those functions in $C_0[0, 1]$, which are somewhere differentiable, form a subset of $C_0[0, 1]$ of the first category, proving the existence of nowhere differentiable functions. Note that we can extend the functions in $C_0[0, 1]$ to the whole of $\mathbb{R}$ by periodicity and consider the space $\Gamma$ of all such functions with the metric as defined above in (5.7).

Let $\mathcal{K} \subset \Gamma$ be the set of all functions $f$ such that, for some $\xi$, the set

$$\left\{ \frac{f(\xi + h) - f(\xi)}{h}, \ h > 0 \right\}$$

is bounded. It is clear that $\mathcal{K}$ contains the set of functions that are somewhere differentiable. We claim that $\mathcal{K}$ is of the first category. Let

$$K_n = \left\{ f \in \Gamma / \text{ for some } \xi, \ \left| \frac{f(\xi + h) - f(\xi)}{h} \right| \leq n, \ \text{for all } h > 0 \right\},$$

$n = 1, 2, \ldots$. Then $\mathcal{K} = \bigcup_{n=1}^{\infty} K_n$. We next claim that $K_n$ is closed, $n = 1, 2, \ldots$. Now, let $f$ be a limit point of $K_n$. Let $\{f_k\}$ be a sequence in $K_n$ converging to $f$. For each $k = 1, 2, \ldots$, let $\xi_k \in [0, 1]$ such that

$$\left| \frac{f_k(\xi_k + h) - f_k(\xi_k)}{h} \right| \leq n, \ \text{for all } h > 0.$$

$\{\xi_k\}$ is a sequence of points in $[0, 1]$ and so $\{\xi_k\}$ is bounded. It has a limit point $\xi$, in view of the Bolzano-Weierstrass theorem. Hence there exists a subsequence $\{\xi_{k_i}\}$ of $\{\xi_k\}$ such that $\xi_{k_i} \to \xi$, $i \to \infty$. Without loss of generality, we can suppose that $\xi_k \to \xi$, $k \to \infty$. For $h > 0$, note that

$$\left|\frac{f(\xi+h)-f(\xi)}{h}\right| \leq \left|\frac{f_k(\xi_k+h)-f_k(\xi_k)}{h}\right| + \frac{1}{h}\left[\begin{array}{l} |f(\xi+h)-f(\xi_k+h)| \\ +|f(\xi_k+h)-f_k(\xi_k+h)| \\ +|f_k(\xi_k)-f(\xi_k)| \\ +|f(\xi_k)-f(\xi)| \end{array}\right].$$

Now, for any $\epsilon > 0$, there exists a positive integer $N = N(\epsilon, h)$ such that

$$\sup_{t\in[0,1]} |f_k(t)-f(t)| < \frac{\epsilon h}{4}, \quad \text{for all } k \geq N.$$

Using the continuity of $f$, there exists a positive integer $M > N$ such that, for all $k \geq M$, we have

$$|f(\xi+h)-f(\xi_k+h)| < \frac{\epsilon h}{4}$$

and

$$|f(\xi_k)-f(\xi)| < \frac{\epsilon h}{4}, \text{ since } \lim_{k\to\infty} \xi_k = \xi.$$

Thus, for all $k \geq M$, we have

$$\begin{aligned} \left|\frac{f(\xi+h)-f(\xi)}{h}\right| &< \left|\frac{f_k(\xi_k+h)-f_k(\xi_k)}{h}\right| + \epsilon \\ &\leq n+\epsilon, \quad \text{for all } \epsilon > 0. \end{aligned}$$

Consequently,

$$\left|\frac{f(\xi+h)-f(\xi)}{h}\right| \leq n, \quad \text{for all } h > 0,$$

and so $f \in K_n$, proving our claim that $K_n$ is closed, $n = 1, 2, \ldots$. We now prove that $\Gamma - K_n$ is dense, i.e., $(\overline{K_n'}) = X$, $n = 1, 2, \ldots$. To this end, let $g \in \Gamma$ and $\epsilon > 0$. We can partition $[0,1]$ into $k$ equal subintervals such that, if $x, x'$ are in the same interval of the partition, then,

$$|g(x)-g(x')| < \frac{\epsilon}{2}.$$

For $x \in [\frac{i-1}{k}, \frac{i}{k}]$, consider the rectangle

$$g\left(\frac{i-1}{k}\right) - \frac{\epsilon}{2} \leq y \leq g\left(\frac{i-1}{k}\right) + \frac{\epsilon}{2}.$$

Then by joining the points $(\frac{i-1}{k}, g(\frac{i-1}{k}))$ and $(\frac{i}{k}, g(\frac{i}{k}))$ by a polygonal path that remains within the rectangle and whose line segments have slopes exceeding $n$ in absolute value, we get a continuous function within the $\epsilon$ neighborhood of $g$ that is in $\Gamma - K_n$. Hence $\Gamma - K_n$ is dense. From what we proved above, $K_n$ is nowhere dense, $n = 1, 2, \ldots$. So $\mathcal{K}$ is of the first category. Since any subset of a set of the first category is again a set of the first category, it follows that the set of functions that are somewhere differentiable, is of the first category. Consequently, the set of functions, which are nowhere differentiable, is of the second category. We know that a set of the second category is not empty. Hence there exist continuous functions on $[a, b]$ that are nowhere differentiable on $[a, b]$, proving our contention. □

**Remark 2.** *Theorem 43 implies that the class of functions that are everywhere continuous but nowhere differentiable on $[a, b]$ is large enough.*

---

## 5.2 Different types of convergence

**Definition 51.** *Let $X$ be a normed linear space. A sequence $\{x_n\}$ in $X$ is said to converge weakly to $x \in X$, written as $x_n \xrightarrow{w} x$, $n \to \infty$, if*

$$f(x_n) \to f(x), \ n \to \infty, \quad \text{for every } f \in \tilde{X}.$$

*$x$ is called the "weak limit" of $\{x_n\}$.*

**Note 8.** *The weak limit, when it exists, is unique, i.e., if $x_n \xrightarrow{w} x$, $x_n \xrightarrow{w} y$, $n \to \infty$, $x, y \in X$, then $x = y$. If $x_n \xrightarrow{w} x$, $x_n \xrightarrow{w} y$, $n \to \infty$, then*

$$f(x_n) \to f(x), \ f(x_n) \to f(y), \ n \to \infty.$$

*Thus, $f(x) = f(y)$, for every $f \in \tilde{X}$, i.e., $f(x - y) = 0$ for all $f \in \tilde{X}$. It now follows that $x - y = 0$, i.e., $x = y$, using the fact that if all bounded linear functionals vanish on a vector, then the vector must be 0.*

**Note 9.** *It is easily proved that if $x_n \overset{w}{\to} x$, $y_n \overset{w}{\to} y$, $n \to \infty$ and $\alpha$ is a scalar, then*

$$x_n + y_n \overset{w}{\to} x + y, \ n \to \infty$$

*and*

$$\alpha x_n \overset{w}{\to} \alpha x, \ n \to \infty.$$

In this context, the usual convergence is called "strong convergence."

**Note 10.** *In a normed linear space, strong convergence always implies weak convergence. However, the converse is not true. The reader can justify the claim by providing a suitable counterexample.*

**Theorem 44.** *In a finite-dimensional space $X$, strong convergence is equivalent to weak convergence.*

*Proof.* It suffices to prove that weak convergence implies strong convergence. Let $X$ be finite-dimensional and $\{e_1, e_2, \ldots, e_k\}$ be a basis of $X$. Let $x_n \overset{w}{\to} x$, $n \to \infty$,

$$x_n = \alpha_1^{(n)} e_1 + \alpha_2^{(n)} e_2 + \cdots \alpha_k^{(n)} e_k, \ n = 1, 2, \ldots$$

and

$$x = \alpha_1 e_1 + \alpha_2 e_2 + \cdots + \alpha_k e_k,$$

where the $\alpha$'s are scalars. Consider the linear functionals $f_i \in \tilde{X}$, $i = 1, 2, \ldots, k$ defined by

$$f_i(e_j) = \delta_{ij},$$

where $\delta_{ij}$ is the Kronecker delta, i.e., $\delta_{ij} = 1$, if $j = i$ and $\delta_{ij} = 0$, if $j \neq i$. Since $x_n \overset{w}{\to} x$, $n \to \infty$,

$$f_i(x_n) \to f_i(x), \ n \to \infty, \ \text{for every } i = 1, 2, \ldots, k.$$

Using the definition of $f_i$,

$$f_i(x_n) = \alpha_i^{(n)} \text{ and } f_i(x) = \alpha_i$$

so that

$$\alpha_i^{(n)} \to \alpha_i,\ n \to \infty, \quad \text{for every } i = 1, 2, \ldots, k.$$

Now, for any $\epsilon > 0$, there exists a positive integer $N$ such that

$$\|\alpha_i^{(n)} - \alpha_i\| < \frac{\epsilon}{kM}, \quad \text{for all } n \geq N \text{ and for all } i = 1, 2, \ldots, k,$$

where $M = \max\limits_{1 \leq i \leq k} \|e_i\|$. Thus,

$$\begin{aligned}
\|x_n - x\| &= \left\| \sum_{i=1}^{k} (\alpha_i^{(n)} - \alpha_i) e_i \right\| \\
&\leq \sum_{i=1}^{k} |\alpha_i^{(n)} - \alpha_i| \|e_i\| \\
&< Mk \frac{\epsilon}{kM} \\
&= \epsilon, \text{ for all } n \geq N.
\end{aligned}$$

It now follows that $x_n \to x$, $n \to \infty$ in the strong sense, completing the proof of the theorem. □

**Theorem 45.** *Let* $\{x_n\}$ *be a bounded sequence in a normed linear space* $X$. *Let* $A$ *be fundamental in* $\tilde{X}$, *i.e.,* $[\overline{A}] = \tilde{X}$. *For* $x \in X$, *let*

$$f(x_n) \to f(x),\ n \to \infty,$$

*for all bounded linear functionals* $f \in A$. *Then*

$$x_n \xrightarrow{w} x,\ n \to \infty.$$

*Proof.* Let $h \in \tilde{X} = [\overline{A}]$. We will prove that

$$h(x_n) \to h(x),\ n \to \infty.$$

In the first instance, let $h \in [A]$. Then

$$h = \sum_{i=1}^{k} \alpha_i f_i, \text{ where } \alpha_i' s \text{ are scalars and } f_i \in A, i = 1, 2, \ldots, k.$$

Now,

$$h(x_n) = \sum_{i=1}^{k} \alpha_i f_i(x_n) \to \sum_{i=1}^{k} \alpha_i f_i(x) = h(x), \ n \to \infty,$$

noting that

$$f_i(x_n) \to f_i(x), \ n \to \infty, \ i = 1, 2, \ldots, k.$$

We are through in this case. Next, let $h \in \overline{[A]}$. So, given $\epsilon > 0$, there exists $\{g_j\}$ in $[A]$ and a positive integer $N$ such that

$$\|h - g_j\| < \epsilon, \text{ for all } j \geq N.$$

By hypothesis, there exists $M > 0$ such that

$$\|x_n\| \leq M, \text{ for all } n = 1, 2, \ldots$$

and

$$\|x\| \leq M.$$

Now, for $j \geq N$,

$$\begin{aligned}
|h(x_n) - h(x)| &\leq |h(x_n) - g_j(x_n)| + |g_j(x_n) - g_j(x)| + |g_j(x) - h(x)| \\
&= |(h - g_j)(x_n)| + |g_j(x_n) - g_j(x)| + |(g_j - h)(x)| \\
&\leq \|(h - g_j)\|\|x_n\| + |g_j(x_n) - g_j(x)| + \|(g_j - h)\|\|x\| \\
&< \epsilon M + |g_j(x_n) - g_j(x)| + \epsilon M.
\end{aligned}$$

Since $g_j \in [A]$,

$$|g_j(x_n) - g_j(x)|$$

can be made arbitrarily small, using the first part of the proof. Hence,

$$h(x_n) \to h(x), \ n \to \infty.$$

It now follows that

$$x_n \overset{w}{\to} x, \ n \to \infty,$$

completing the proof. □

We now discuss convergence in $B(X,Y)$, where $X, Y$ denote normed linear spaces.

**Definition 52.** *A sequence $\{A_n\}$ in $B(X,Y)$ is said to converge "uniformly" to $A \in B(X,Y)$ if, for any $\epsilon > 0$, there exists a positive integer $N$ such that*

$$\|A_n - A\| < \epsilon, \text{ for all } n \geq N.$$

In the present case, we define strong convergence as follows:

**Definition 53.** *A sequence $\{A_n\}$ in $B(X,Y)$ is said to converge "strongly" to $A \in B(X,Y)$, written as $A_n \xrightarrow{s} A$, $n \to \infty$, if $A_n(x) \to A(x)$, $n \to \infty$ for any $x \in X$.*

**Definition 54.** *A sequence $\{A_n\}$ in $B(X,Y)$ is said to converge "weakly" to $A \in B(X,Y)$, written as $A_n \xrightarrow{w} A$, $n \to \infty$, if for any $x \in X$, the sequence $\{A_n(x)\}$ in $Y$ converges weakly to $A(x) \in Y$, i.e.,*

$$f(A_n(x)) \to f(A(x)), \ n \to \infty, \text{ for any } f \in \tilde{Y}.$$

**Note 11.** *In $B(X,Y)$, uniform convergence implies strong convergence, which implies weak convergence.*

*Proof.* Let $\{A_n\}$ be uniformly convergent to $A$ in $B(X,Y)$, i.e.,

$$\|A_n - A\| \to 0, \ n \to \infty.$$

Now, for any $x \in X$,

$$\begin{aligned} \|A_n(x) - A(x)\| &= \|(A_n - A)(x)\| \\ &\leq \|A_n - A\|\|x\| \\ &\to 0, \ n \to \infty, \end{aligned}$$

so that $A_n(x) \to A(x)$, $n \to \infty$, for any $x \in X$, i.e., $A_n \xrightarrow{s} A$, $n \to \infty$, i.e., uniform convergence implies strong convergence. The remaining part is straightforward. □

**Note 12.** *In particular, strong and weak convergence are equivalent in $\tilde{X}$.*

*Proof.* Let $\{f_n\}$ be a sequence in $\tilde{X}$. Then

$$f_n \xrightarrow{s} f, n \to \infty, f \in \tilde{X}$$

if and only if

$$f_n(x) \xrightarrow{s} f(x), n \to \infty, \text{ for any } x \in X \tag{5.8}$$

and

$$f_n \xrightarrow{w} f, n \to \infty$$

if and only if

$$f_n(x) \xrightarrow{w} f(x), n \to \infty, \text{ for any } x \in X. \tag{5.9}$$

But the convergences (5.8) and (5.9) occur in the underlying field $\mathbb{R}$ or $\mathbb{C}$, which is a one-dimensional linear space and hence they are equivalent in view of Theorem 44. Thus strong and weak convergence in $\tilde{X}$ are equivalent. □

**Remark 3.** *We now wish to make the terminology in $\tilde{X}$ clear. In view of the equivalence of the strong and weak convergence in $\tilde{X}$, we drop the name strong convergence in $\tilde{X}$ and call it weak convergence. The convergence with respect to the norm in $\tilde{X}$, which is called uniform convergence, is called strong convergence in $\tilde{X}$. So, we say that the sequence $\{f_n\}$ in $\tilde{X}$ is strongly convergent to $f$ in $\tilde{X}$ if*

$$\|f_n - f\| \to 0, \ n \to \infty.$$

**Definition 55.** *Let*

$$J : X \to \tilde{\tilde{X}}$$
$$x \to x''$$

*be the mapping defined by*

$$x''(x') = x'(x), \ \textit{where } x' \in \tilde{X}.$$

*A sequence $\{f_n\}$ is said to be "weak*" convergent to $f$ in $B(X,Y)$, written as*

$$f_n \overset{w^*}{\rightarrow} f, \; n \to \infty$$

*if, for all $x \in X$,*

$$(J(x))(f_n) \to (J(x))(f), \; n \to \infty.$$

**Theorem 46.** *The sequence $\{A_n\}$ in $B(X,Y)$ converges uniformly to $A$ in $B(X,Y)$ if and only if $A_n(x) \to A(x)$, $n \to \infty$ uniformly for all $x \in X$ with $\|x\| \leq 1$.*

*Proof.* Let $A_n \to A$, $n \to \infty$ uniformly. Then, for $\epsilon > 0$, there exists a positive integer $N$ such that

$$\|A_n - A\| < \epsilon, \text{ for all } n \geq N.$$

If $\|x\| \leq 1$,

$$\begin{aligned} \|A_n(x) - A(x)\| &= \|(A_n - A)(x)\| \\ &\leq \|A_n - A\| \|x\| \\ &\leq \|A_n - A\|, \text{ since } \|x\| \leq 1 \\ &< \epsilon, \text{ for all } n \geq N. \end{aligned}$$

It is clear that $N$ does not depend on the choice of $x$ with $\|x\| \leq 1$. Thus, uniform convergence follows.

Conversely, suppose that for $\epsilon > 0$ there exists a positive integer $N$, depending only on $\epsilon$, such that

$$\|A_n(x) - A(x)\| < \epsilon, \text{ for all } n \geq N \text{ and for all } x \text{ with } \|x\| \leq 1.$$

Let $x \in X$, $x \neq 0$. Let $y = \frac{x}{\|x\|}$. Then $\|y\| = 1$. It is clear that

$$\begin{aligned} &\|A_n(y) - A(y)\| < \epsilon, \\ &\textit{i.e.}, \|A_n(x) - A(x)\| < \epsilon\|x\|, \text{ for any } x \neq 0, \end{aligned}$$

which holds even if $x = 0$. So the above inequality holds for all $x$ and so

$$\|A_n - A\| < \epsilon, \text{ for all } n \geq N,$$

which implies that $A_n \to A$, $n \to \infty$ uniformly. The proof is now complete. □

We now characterize weak convergence in $\ell_p$, $1 < p < \infty$.

**Theorem 47.** *Let* $1 < p < \infty$ *and*

$$x_n = \{\alpha_1^{(n)}, \alpha_2^{(n)}, \dots, \alpha_k^{(n)}, \dots\} \in \ell_p, \ n = 1, 2, \dots$$

*and*

$$x = \{\alpha_1, \alpha_2, \dots, \alpha_k, \dots\} \in \ell_p.$$

*Then*

$$x_n \overset{w}{\to} x, \ n \to \infty$$

*if and only if*

*(i)* $\{x_n\}$ *is bounded, i.e.,* $\|x_n\| \leq M$, $M > 0$, $n = 1, 2, \dots$;

*and*

*(ii) for every* $i = 1, 2, \dots$, $\alpha_i^{(n)} \to \alpha_i$, $n \to \infty$.

*Proof.* Sufficiency. Let $y = \{\xi_n\} \in \ell_q$, $\frac{1}{p} + \frac{1}{q} = 1$. Then, we can write

$$y = \sum_{n=1}^{\infty} \xi_n e_n,$$

with the usual meaning of $e_n$, $n = 1, 2, \dots$. Since $\tilde{\ell}_p$ is congruent to $\ell_q$, corresponding to $e_i$, there exists $f_i$, $\in \tilde{\ell}_p$, $i = 1, 2, \dots$ under the congruence mapping. We know that we can also write $f \in \tilde{\ell}_p$ as

$$f = \sum_{i=1}^{\infty} \beta_i f_i, \text{ where } \beta_i = f(e_i), \ i = 1, 2, \dots.$$

Let $A = \{f_1, f_2, \dots, f_n, \dots\}$. Then $[\overline{A}] = \tilde{\ell}_p$. We also have

$$f_i(x_n) = \alpha_i^{(n)} \text{ and } f_i(x) = \alpha_i.$$

By hypothesis,

$$\alpha_i^{(n)} \to \alpha_i, \; n \to \infty, \; i = 1, 2, \ldots .$$

Thus,

$$f_i(x_n) \to f_i(x), \; n \to \infty, \; i = 1, 2, \ldots,$$

which implies that $x_n \overset{w}{\to} x$, $n \to \infty$.

Necessity. Let $x_n \overset{w}{\to} x$, $n \to \infty$. Then $\{x_n\}$ is bounded, i.e., there exists $M > 0$ such that

$$\|x_n\| \leq M, \; n = 1, 2, \ldots .$$

(The above result will be proved later after the Banach–Steinhaus theorem). Since $x_n \overset{w}{\to} x$, $n \to \infty$,

$$f(x_n) \to f(x), \; n \to \infty, \; \text{for every } f \in \tilde{\ell}_p.$$

Consequently,

$$f_i(x_n) \to f_i(x), \; n \to \infty, \; i = 1, 2, \ldots,$$
$$i.e., \alpha_i^{(n)} \to \alpha_i, \; n \to \infty, \; i = 1, 2, \ldots,$$

completing the proof of the theorem. □

---

## 5.3 Principle of uniform boundedness (Banach–Steinhaus theorem)

In this section, we prove the uniform boundedness theorem, which is also known as the Banach–Steinhaus theorem.

**Theorem 48** (Banach–Steinhaus)**.** *Let $\Lambda$ be an indexing set of any cardinality. Let $\{A_\alpha / \alpha \in \Lambda\}$ be a collection of elements of $B(X, Y)$, where $X$ is a Banach space. If*

$$\sup_{\alpha \in \Lambda} \|A_\alpha(x)\| < \infty, \; \textit{for all } x \in X,$$

*then*

$$\sup_{\alpha \in \Lambda} \|A_\alpha\| < \infty,$$

*i.e., there exists $M > 0$ such that $\|A_\alpha\| \leq M$, for all $\alpha \in \Lambda$, i.e., the set of all real numbers $\{\|A_\alpha\|, \alpha \in \Lambda\}$ is uniformly bounded by $M$.*

*Proof.* Case 1. Let $\Lambda$ be denumerable or countable. Let $\Lambda = \{A_1, A_2, \ldots, A_n, \ldots\}$ be a countable set of elements of $B(X, Y)$ satisfying the hypothesis of the theorem. We now make two claims.

Claim 1. For some $x_0 \in X$ and $\delta > 0$, there exists $k > 0$ such that

$$\|A_n(x)\| \leq k, \text{ for all } n = 1, 2, \ldots \text{ and for all } x \in \overline{S_\delta(x_0)}.$$

Claim 2. Claim 1 indeed implies the theorem.

We first prove claim 2 using claim 1 and then prove claim 1. We shall suppose that claim 1 is fulfilled. Let $x \in X$ be such that $\|x\| \leq \delta$. Then

$$\|A_n(x)\| \leq \|A_n(x + x_0)\| + \|A_n(x_0)\|.$$

Since

$$\|(x + x_0) - x_0\| = \|x\| \leq \delta,$$

$x + x_0 \in S_\delta(x_0)$ and so, using claim 1,

$$\|A_n(x + x_0)\| \leq k, \text{ for all } n = 1, 2, \ldots.$$

Since $\|A_n(x_0)\| \leq k$, it follows that

$$\begin{aligned} \|A_n(x)\| &\leq k + k \\ &= 2k, \text{ for all } x \in X \text{ such that } \|x\| \leq \delta. \end{aligned}$$

Now, let $x \in X$, $x \neq 0$. Consider $\frac{\delta x}{\|x\|}$. Then $\left\|\frac{\delta x}{\|x\|}\right\| = \delta$.

$$\begin{aligned} \|A_n(x)\| &= \left\| \frac{\|x\|}{\delta} A_n\left(\frac{\delta x}{\|x\|}\right) \right\| \\ &\leq \frac{\|x\|}{\delta} 2k, \text{ for all } n = 1, 2, \ldots. \end{aligned}$$

The above inequality holds for $x = 0$ too. Thus,

$$\|A_n(x)\| \leq \frac{2k}{\delta}\|x\|, \text{ for all } x \in X, n = 1, 2, \ldots.$$

This implies that

$$\|A_n\| \leq \frac{2k}{\delta}, \text{ for all } n = 1, 2, \ldots,$$

$$i.e., \sup_n \|A_n\| \leq \frac{2k}{\delta},$$

proving the theorem in this case. To complete the proof, we are yet to prove claim 1. Suppose claim 1 does not hold, i.e., let us suppose that there exists no $k > 0$, satisfying the requirements mentioned in claim 1. Thus, for any (spherical) neighborhood $S$ of any point and any real number $a > 0$, there exists a positive integer $n_0$ and a point $x_0 \in S$ such that

$$\|A_{n_0}(x_0)\| > a.$$

Hence, there exists a neighborhood $S_a$ of $x_0$ such that

$$\overline{S}_a \subset S \text{ and } \|A_{n_0}(x)\| > a, \text{ for all } x \in \overline{S}_a.$$

Taking $a = 1$, we can now assert the existence of a closed set $S_1$ such that

$$S_1 \subset S \text{ and } \|A_{n_1}(x)\| > 1, \text{ for all } x \in S_1,$$

where we can assume without loss of generality that

$$d(S_1) < 1, \; d(S_1) = \sup_{a,b \in S_1} d(a, b).$$

Proceeding similarly, we can find a closed set $S_2 \subset S_1$ and a positive integer $n_2$ such that

$$\|A_{n_2}(x)\| > 2, \text{ for all } x \in S_2 \text{ and } d(S_2) < \frac{1}{2}.$$

By the above procedure, we generate a nested sequence of closed sets, whose diameters tend to zero. Hence there exists $y \in \bigcap_{n=1}^{\infty} S_n$, in view of the nested sequence theorem. This implies that

$$\|A_{n_k}(y)\| > k, \text{ for all } k = 1, 2, \ldots,$$

which contradicts the hypothesis that $\sup_n \|A_n(x)\| < \infty$, for all $x \in X$, establishing claim 1.

Case 2. We shall now prove the theorem if $\Lambda$ is any arbitrary indexing set. It is clear that if $\{A_\alpha/\alpha \in \Lambda\}$ satisfies the hypothesis, then any countable subset of $\Lambda$ also satisfies the hypothesis. If possible, let

$$\sup_{\alpha\in\Lambda} \|A_\alpha\| = \infty.$$

Then, we can get $A_1, A_2, \ldots, A_n, \ldots$ such that

$$\|A_n\| > n, \ n = 1, 2, \ldots. \tag{5.10}$$

However, $\{A_1, A_2, \ldots, A_n, \ldots\}$ is countable and by case 1,

$$\sup_n \|A_n\| < \infty,$$

which contradicts (5.10), proving the theorem in this case too. The proof of the theorem is now complete. □

**Corollary 2.** *Let $A$ be a subset of a normed linear space such that*

$$\{x'(x)/x \in A\} \text{ is bounded, for all } x' \in \tilde{X}.$$

*Then $A$ is a bounded subset of $X$.*

*Proof.* Consider the mapping

$$J : X \to \tilde{\tilde{X}}$$
$$x \to x'',$$

defined by

$$x''(x') = x'(x).$$

Since

$$\{x'(x)/x \in A\} \text{ is bounded, for all } x' \in \tilde{X},$$
$$\{x''(x')/x \in A\} \text{ is bounded, for all } x' \in \tilde{X}.$$

Since $\tilde{X}$ is a Banach space, in view of Banach–Steinhaus theorem, the set

$$\{x''/x \in A\} \text{ is bounded.}$$

Since $\|x''\| = \|x\|$, it follows that $A$ is bounded, completing the proof. □

## 5.4 Some consequences and applications of the Banach–Steinhaus theorem

**Theorem 49.** *Let $\{A_n\}$ be a sequence in $B(X, Y)$, where $X$ is a Banach space, converging strongly to $A$ in $B(X, Y)$. Then, there exists $M > 0$ such that $\|A_n\| \leq M$, for all $n = 1, 2, \ldots$.*

*Proof.* Since $A_n \xrightarrow{s} A$, $n \to \infty$, $A_n(x) \to A(x)$, $n \to \infty$, for all $x \in X$. Hence,

$$\sup_n \|A_n(x)\| < \infty, \text{ for all } x \in X.$$

Using the Banach–Steinhaus theorem, we have

$$\sup_n \|A_n\| < \infty,$$

which is what we desired. □

**Definition 56.** *A sequence $\{A_n\}$ in $B(X, Y)$ is called a "strong Cauchy sequence," if $\{A_n(x)\}$ is a Cauchy sequence in $Y$, for all $x \in X$.*

**Definition 57.** *We say that $B(X, Y)$ is "complete in the strong sense," if every strong Cauchy sequence of vectors in $B(X, Y)$ converges strongly to some element of $B(X, Y)$.*

**Theorem 50.** *If $X, Y$ are Banach spaces, then $B(X, Y)$ is complete in the strong sense.*

*Proof.* Let $\{A_n\}$ be a strong Cauchy sequence in $B(X, Y)$. Then, for all $x \in X$, $\{A_n(x)\}$ is a Cauchy sequence in $Y$. Since $Y$ is a Banach space, this sequence is convergent. Let

$$\lim_{n\to\infty} A_n(x) = A(x).$$

It is clear that $A$ is linear and $A_n \xrightarrow{s} A$, $n \to \infty$. Since $X$ is a Banach space, using Theorem 49,

$$\|A_n\| \leq M, \; M > 0, \; n = 1, 2, \ldots.$$

Now, for any $x \in X$,

$$\begin{aligned}\|A_n(x)\| &\leq \|A_n\|\|x\| \\ &\leq M\|x\|, \text{ for all } n = 1, 2, \ldots.\end{aligned}$$

Consequently,

$$\begin{aligned}\lim_{n\to\infty} \|A_n(x)\| &\leq M\|x\|, \\ i.e., \left\|\lim_{n\to\infty} A_n(x)\right\| &\leq M\|x\|, \\ i.e., \|A(x)\| &\leq M\|x\|, \text{ for all } x \in X,\end{aligned}$$

which implies that $A$ is bounded, i.e., $A \in B(X,Y)$, thus completing the proof. □

**Definition 58.** *The sequence $\{x_n\}$ in a normed linear space $X$ is called a "weak Cauchy sequence," if $\{f(x_n)\}$ is a Cauchy sequence, for all $f \in \tilde{X}$.*

We now prove an important result, which we have already used.

**Theorem 51.** *Every weak Cauchy sequence in a normed linear space is bounded.*

*Proof.* Let $\{x_n\}$ be a weak Cauchy sequence in $X$, i.e., $\{f(x_n)\}$ is a Cauchy sequence, for all $f \in \tilde{X}$. Let, as usual,

$$\begin{aligned}J : X &\to \tilde{\tilde{X}} \\ x &\to J(x)\end{aligned}$$

be the mapping defined by

$$(J(x))(y) = y(x), \text{ for every } y \in \tilde{X}.$$

Then we have

$$\|J(x)\| = \|x\|.$$

Since $\{f(x_n)\}$ is a Cauchy sequence of scalars (real or complex numbers), we have, for any $f \in \tilde{X}$,

$$\sup_n |J(x_n)(f)| = \sup_n |f(x_n)| < \infty.$$

Since $\tilde{X}$ is a Banach space, using the Banach–Steinhaus theorem, it follows that

$$\sup_n \|J(x_n)\| < \infty,$$

$$i.e., \sup_n \|x_n\| < \infty, \ \text{since } \|J(x_n)\| = \|x_n\|, \ n = 1, 2, \ldots,$$

proving the result. □

Noting that a weak convergent sequence is a weak Cauchy sequence, we have:

**Corollary 3.** *In the normed linear space $X$, if the sequence $\{x_n\}$ converges weakly, then it is bounded.*

**Definition 59.** *A normed linear space $X$ is said to be "weakly complete" if every weak Cauchy sequence in $X$ converges weakly to some element of $X$.*

**Theorem 52.** *If the normed linear space $X$ is reflexive, it is weakly complete.*

*Proof.* Let $\{x_n\}$ be a weak Cauchy sequence in $X$, i.e., $\{f(x_n)\}$ is a Cauchy sequence, for every $f \in \tilde{X}$. As usual, consider the mapping

$$J : X \to \tilde{\tilde{X}}$$

defined by

$$(J(x))(y) = y(x), \ \text{for every } y \in \tilde{X}.$$

Since

$$(J(x_n))(f) = f(x_n), n = 1, 2, \ldots,$$

$\{J(x_n)(f)\}$ is a Cauchy sequence of scalars, for every $f \in \tilde{X}$. Since $\mathbb{R}$ or $\mathbb{C}$ is complete, $\lim_{n\to\infty} (J(x_n))(f)$ exists, for every $f \in \tilde{X}$. Let

$$\lim_{n\to\infty} (J(x_n))(f) = y(f), \ \text{for every } f \in \tilde{X}.$$

Clearly, $y$ is a linear functional on $\tilde{X}$. We now claim that $y$ is bounded. Since

$$\|J(x_n)\| = \|x_n\|, \ n = 1, 2, \ldots,$$

and $\{x_n\}$ is a Cauchy sequence, $\{x_n\}$ is bounded, i.e., there exists $M > 0$ such that

$$\|x_n\| \leq M, \ n = 1, 2, \ldots.$$

Thus,

$$|(J(x_n))(f)| \leq M\|f\|, \text{ for all } f \in \tilde{X}, \ n = 1, 2, \ldots.$$

Consequently,

$$|y(f)| \leq M\|f\|, \text{ for all } f \in \tilde{X},$$

$$i.e., y \text{ is bounded},$$

and so $y \in \tilde{\tilde{X}}$. Since $X$ is reflexive, there exists $x \in X$ such that

$$y = J(x).$$

Hence, for any $f \in \tilde{X}$, we have

$$\lim_{n\to\infty} f(x_n) = \lim_{n\to\infty} (J(x_n))(f) = y(f) = (J(x))(f) = f(x).$$

So, for any $f \in \tilde{X}$,

$$\lim_{n\to\infty} f(x_n) = f(x),$$

$$i.e., x_n \xrightarrow{w} x, \ n \to \infty,$$

completing the proof. □

**Remark 4.** *We have already noted that every Hilbert space is reflexive and so every Hilbert space is weakly complete, in view of Theorem 52.*

We now consider some applications of the Banach–Steinhaus theorem.

**Theorem 53.** *Let $\{\alpha_n\}$ be a sequence of complex numbers and $p \geq 1$. Let $\sum_{n=1}^{\infty} \alpha_n \xi_n$ converge for every sequence $\{\xi_n\} \in \ell_p$. If $p = 1$, then $\{\alpha_n\} \in \ell_\infty$ and if $p > 1$, $\{\alpha_n\} \in \ell_q$, where $\frac{1}{p} + \frac{1}{q} = 1$.*

*Proof.* Let $x = \{\xi_n\} \in \ell_p$. For $k = 1, 2, \ldots$, let

$$f_k(x) = \sum_{n=1}^{k} \alpha_n \xi_n.$$

It is clear that $f_k$ is a bounded linear functional on $\ell_p$ and so

$$f_k \in \tilde{\ell_p},\ k = 1, 2, \ldots.$$

Also,

$$\|f_k\| = \begin{cases} \max_{1\le i\le k} |\alpha_i|, & \text{if } p = 1; \\ \left(\sum_{i=1}^{k} |\alpha_i|^q\right)^{\frac{1}{q}}, & \text{if } p > 1. \end{cases}$$

Since the infinite series $\sum_{n=1}^{\infty} \alpha_n \xi_n$ converges, for every $x = \{\xi_n\} \in \ell_p$, the set

$$\{f_k(x)/k = 1, 2, \ldots\} \text{ is bounded, for every } x \in \ell_p.$$

In view of the Banach–Steinhaus theorem, the set

$$\{f_k/k = 1, 2, \ldots\}$$

is bounded. It now follows that $\{\alpha_n\} \in \ell_\infty$, if $p = 1$ and $\{\alpha_n\} \in \ell_q$, if $p > 1$. The proof of the result is now complete. □

**Theorem 54.** *There are continuous functions of period $2\pi$, whose Fourier series diverge at a given point.*

*Proof.* Let $X$ be the Banach space of all real-valued, continuous functions of period $2\pi$ with

$$\|x\| = \sup |x(t)|,\ x \in X.$$

The $n$th partial sum of the Fourier series of $x = x(t)$ at $t = 0$ is written in the form

$$f_n(x) = \frac{1}{2\pi} \int_0^{2\pi} x(t) D_n(t) dt,$$

where

$$D_n(t) = \frac{\sin\{(n + \frac{1}{2})t\}}{\sin(\frac{1}{2}t)},\ n = 1, 2, \ldots.$$

For every $x \in X$,

$$|f_n(x)| \leq \left\{ \frac{1}{2\pi} \int_0^{2\pi} |D_n(t)|dt \right\} \|x\|.$$

Note that $f_n$ is linear and bounded. Also,

$$\|f_n\| \leq \frac{1}{2\pi} \int_0^{2\pi} |D_n(t)|dt, \ n = 1, 2, \ldots.$$

We now claim that there exists $x \in X$, whose Fourier series fails to converge at $t = 0$. Suppose not. Then, for every $x \in X$, the Fourier series of $x$ converges at $t = 0$, i.e., $\{f_n(x)\}$ converges, for all $x \in X$. So, $\{f_n(x)\}$ is bounded for all $x \in X$. In view of the Banach–Steinhaus theorem, the sequence $\{\|f_n\|\}$ is bounded. We shall arrive at a contradiction, by first proving that

$$\|f_n\| = \frac{1}{2\pi} \int_0^{2\pi} |D_n(t)|dt, \ n = 1, 2, \ldots$$

and then proving that $\left\{ \int_0^{2\pi} |D_n(t)|dt \right\}$ diverges. Let $y$ be the function, which is $+1$ at those $t$ for which $D_n(t) \geq 0$ and $-1$ at those $t$ for which $D_n(t) < 0$. For every $\epsilon > 0$, $y$ may be modified to a continuous function $x$ of norm 1 such that

$$\begin{aligned} \left| f_n(x) - \frac{1}{2\pi} \int_0^{2\pi} |D_n(t)|dt \right| \\ &= \frac{1}{2\pi} \left| \int_0^{2\pi} (x(t) - y(t)) D_n(t)dt \right| \\ &< \epsilon, \end{aligned}$$

from which it follows that

$$\|f_n\| = \frac{1}{2\pi} \int_0^{2\pi} |D_n(t)|dt.$$

We now prove that

$$\left\{ \int_0^{2\pi} |D_n(t)|dt \right\} = \left\{ \int_0^{2\pi} \left| \frac{\sin(n + \frac{1}{2})t}{\sin \frac{1}{2}t} \right| dt \right\}$$

is unbounded. Since $\sin u \leq u$, for $0 \leq u \leq \pi$, it suffices to show that

$$\left\{ \int_0^{\pi} \left| \frac{\sin(2n + 1)u}{u} \right| du \right\}$$

is unbounded. Now,

$$\begin{aligned}\int_0^{\pi} \left|\frac{\sin(2n+1)u}{u}\right| du &= \sum_{k=0}^{2n} \int_{\frac{k\pi}{2n+1}}^{\frac{(k+1)\pi}{2n+1}} \left|\frac{\sin(2n+1)u}{u}\right| du \\ &\geq \sum_{k=0}^{2n} \frac{2n+1}{(k+1)\pi} \int_{\frac{k\pi}{2n+1}}^{\frac{(k+1)\pi}{2n+1}} |\sin(2n+1)u| du \\ &= \sum_{k=0}^{2n} \frac{1}{(k+1)\pi} \int_{k\pi}^{(k+1)\pi} |\sin x| dx \\ &= \frac{2}{\pi} \sum_{k=0}^{2n} \frac{1}{k+1}.\end{aligned}$$

We know that the series $\sum_{n=1}^{\infty} \frac{1}{n}$ diverges and so $\{\|f_n\|\}$ is unbounded, which is a contradiction. Thus, there exists some $x \in X$, whose Fourier series diverges at $t = 0$, completing the proof. □

The Banach–Steinhaus theorem has a similar application to interpolation theory as well (for details, the reader can refer to ([7], pp. 78–79). We shall exhibit more applications of the Banach–Steinhaus theorem in "summability theory" in Chapter 8.

## Exercises

1. Prove that a subset $A$ of a normed linear space $X$ is nowhere dense if and only if $(\overline{A})'$ is dense in $X$, i.e., $\overline{(\overline{A})'} = X$. (Note that this equivalent formulation of a nowhere dense set was used in several places in Chapter 5).

2. Prove that a subset $A$ of a normed linear space $X$ is nowhere dense if and only if every non-empty open set $U$ has a non-empty open subset $V$ such that $V \cap A = \phi$.

3. Prove that if $A$ is any open set in a normed linear space $X$, $\overline{A} - A$ is nowhere dense.

4. Prove that the Cantor set is nowhere dense.

5. If $X$ is a reflexive space, prove that every infinite bounded subset of $X$ contains a weak Cauchy sequence.

6. Prove that a subset $A$ of a normed linear space $X$ is bounded if and only if $f(A)$ is bounded for every $f \in \tilde{X}$.

7. Prove that, in a finite-dimensional linear space $X$, the three concepts of convergence for operators, viz., uniform, strong, and weak, coincide.

8. Does the Banach–Steinhaus theorem hold when $X$ is not complete? Justify your answer.

9. If $X$ is separable, prove that any bounded sequence in $\tilde{X}$ contains a weakly convergent subsequence.

10. If $A_n \in L(X, Y)$, $n = 1, 2, \ldots$, where $X, Y$ are normed linear spaces and if $A_n \xrightarrow{s} A$, $n \to \infty$, prove that $A \in L(X, Y)$.

11. If $\sum_{k=1}^{\infty} a_k x_k$ converges, for all $x = \{x_k\} \in c$, prove that $a = \{a_k\} \in \ell_1$.

12. If $\sum_{k=1}^{\infty} a_k x_k$ converges, whenever $\sum_{k=1}^{\infty} x_k$ converges, prove that $a = \{a_k\} \in BV$, i.e., $\sum_{k=1}^{\infty} |a_k - a_{k+1}| < \infty$ (any element of $BV$ is called a sequence of bounded variation).

13. Prove that $BV$ is a Banach space under the norm

$$\|x\| = |x_1| + \sum_{n=1}^{\infty} |x_n - x_{n+1}|, \; x = \{x_n\} \in BV.$$

14. If $\sum_{k=1}^{\infty} |a_k x_k| < \infty$, whenever $\sum_{k=1}^{\infty} x_k$ converges, prove that $a = \{a_k\} \in \ell_1$.

15. Prove that $c$ is a nowhere dense subspace of $\ell_\infty$.

16. Let $s$ be the set of all sequences of real numbers. For $x = \{x_i\}$, $y = \{y_i\} \in s$, let us define

$$d(x, y) = \sum_{i=1}^{\infty} \frac{1}{2^i} \frac{|\alpha_i - \beta_i|}{1 + |\alpha_i - \beta_i|}.$$

Prove that $(s, d)$ is a complete metric space that is also separable.

17. Construct a function $f$ that is discontinuous at every rational point of $[0, 1]$ and continuous at every irrational point of $[0, 1]$.

18. Is it possible to construct a function $f$ that is continuous at every rational point of $[0, 1]$ and discontinuous at every irrational point of $[0, 1]$? Justify your answer.

19. Prove that a closed set $A$ in a normed linear space $X$ is nowhere dense if and only if $A'$ is everywhere dense.

20. Prove that a proper, closed subspace of a normed linear space is nowhere dense.

# Chapter 6

# Closed Graph Theorem and Open Mapping Theorem

## 6.1 Bounded inverse theorem

**Definition 60.** *Let $D$ be a subset of a set $X$ and $A$ be a mapping of $D$ into a set $Y$. Then the set*

$$G_A = \{(x, A(x))/x \in D\}$$

*is called the "graph of $A$."*

**Definition 61.** *Let $X, Y$ be normed linear spaces and $D$ be a subspace of $X$. Then, the linear transformation $A : D \to Y$ is said to be "closed" if*

$$x_n \to x,\ n \to \infty,\ A(x_n) \to y,\ n \to \infty,\ x_n \in D,\ n = 1, 2, \ldots$$

*imply that*

$$x \in D \text{ and } A(x) = y.$$

**Theorem 55.** *$A$ is a closed, linear transformation if and only if its graph is a closed subspace of $X \times Y$.*

*Proof.* Let $D \subset X$ and $A : D \to Y$ be a closed linear transformation. Let $(x, y)$ be a limit point of $G_A$. Then, there exists a sequence of points $(x_n, A(x_n))$ of

$G_A$, converging to $(x, y)$. This implies

$$\|x_n - x, A(x_n) - y\| \to 0, \; n \to \infty,$$

$$\text{i.e.,} \quad \|x_n - x\| + \|A(x_n) - y\| \to 0, \; n \to \infty,$$

$$\text{i.e.,} \quad x_n \to x, \; A(x_n) \to y, \; n \to \infty.$$

Hence,

$$x \in D \text{ and } y = A(x), \text{ since } A \text{ is closed,}$$

$$\text{i.e.,} \quad (x, y) = (x, A(x)) \in G_A$$

and so $G_A$ is closed. Conversely, let $G_A$ be a closed subspace of $X \times Y$. Let $x_n \to x$ and $A(x_n) \to y$, $n \to \infty$, where $x_n \in D$, $n = 1, 2, \ldots$. We claim that $x \in D$ and $y = A(x)$. Now,

$$(x_n, A(x_n)) \to (x, y) \in \overline{G}_A = G_A, \text{ since } G_A \text{ is closed,}$$

$$\text{i.e.,} \, (x, y) \in G_A,$$

which implies that

$$x \in D \text{ and } y = A(x),$$

by the definition of $G_A$. □

We now investigate the connection between continuous linear transformations and closed linear transformations. We exhibit a discontinuous but closed transformation. Let $X = Y = C[0, 1]$,

$$D = \{x \in X / x' \in C[0, 1]\},$$

where $x'$ denotes the derivative of $x$. Let $A$ be the transformation with domain $D$ defined by

$$A(x) = x', \; x \in D.$$

It is clear that $A$ is linear. If $x_n(t) = t^n$, $n = 1, 2, \ldots$, then $\|x_n\| = 1$ and $x'_n(t) = nt^{n-1}$ and so

$$\|A(x_n)\| = \|x'_n\| = n, \; n = 1, 2, \ldots.$$

Consequently,

$$\sup_{\|x\|=1} \|A(x)\| = \infty,$$

and so $A$ is unbounded. We now claim that $A$ is closed. Let $x_n \in D$, $n = 1, 2, \ldots$, $x_n \to x$, $A(x_n) \to y$, $n \to \infty$. Clearly,

$$x'_n(t) \to y(t), \;\; n \to \infty \text{ uniformly}$$

and $y$ is continuous. It now follows that $x(t)$ is differentiable with $x'(t) = y(t)$. Thus $x \in D$ and $A(x) = y$, proving that $A$ is closed.

**Theorem 56.** *Let $X, Y$ be normed linear spaces and $D$ be a closed subspace of $X$. Let $A : D \to Y$ be a bounded linear transformation. Then $A$ is closed.*

*Proof.* Let $\{x_n\}$ be a convergent sequence of points in $D$, with limit $x$, such that $A(x_n) \to y$, $n \to \infty$. Since $D$ is closed, $x \in D$ and using the continuity of $A$, $A(x_n) \to A(x)$, $n \to \infty$, so that $y = A(x)$. Thus $A$ is closed. □

**Corollary 4.** *Since $X$ is closed, if $A$ is a bounded linear transformation from $X$ into $Y$, then $A$ is closed. Later on, we will prove that the converse of the above result holds using the closed graph theorem.*

**Theorem 57.** *Let $A : D \to Y$ be a closed linear transformation from $D$ into $Y$, where $D$ is a subspace of $X$. If $A^{-1}$ exists, it is also a closed, linear transformation.*

*Proof.* Clearly, the graph of $A^{-1}$ is given by

$$G_{A^{-1}} = \{(Ax, x)/x \in D\}.$$

Consider the mapping

$$X \times Y \to Y \times X$$

$$(x, y) \to (y, x),$$

which is an isometry. Since an isometry maps closed sets into closed sets, $G_{A^{-1}}$ is closed. If $A^{-1}$ exists, clearly it is linear and since the graph of $A^{-1}$ is closed, $A^{-1}$ is closed, proving the theorem. □

**Theorem 58.** *Let $A : D \to Y$ be a linear transformation of $D$ into $Y$, where $D$ is a subspace of $X$ and $Y$ is a Banach space. If $A$ is closed and bounded, then $D$ is a closed subspace of $X$.*

*Proof.* Let $x \in \overline{D}$. Then there exists a sequence $\{x_n\}$ in $D$ such that $x_n \to x$, $n \to \infty$. Now,

$$\begin{aligned}\|A(x_n) - A(x_m)\| &= \|A(x_n - x_m)\| \\ &\leq \|A\|\|x_n - x_m\| \\ &\to 0,\ n, m \to \infty,\end{aligned}$$

since $\{x_n\}$ is a Cauchy sequence. Thus, $\{A(x_n)\}$ is a Cauchy sequence in $Y$ that is complete. So,

$$A(x_n) \to y,\ n \to \infty,\ y \in Y.$$

Since $A$ is closed, it follows that $x \in D$ and $A(x) = y$. Consequently, $\overline{D} = D$, proving that $D$ is closed, proving the theorem. □

We now take up the main theorem of this section.

**Theorem 59** (Bounded inverse theorem). *Let $X$ be a Banach space, $Y$ be a normed linear space, $D$ be a subspace of $X$, $A : D \to Y$ be a closed linear transformation of $D$ into $Y$, and $A(D)$ be of the second category. Then*

*(i) $A$ is onto;*

*(ii) there exists $m > 0$ such that for $y \in Y$, there exists some $x \in D$ such that $A(x) = y$ and $\|x\| \leq m\|y\|$;*

*and*

*(iii) if $A^{-1}$ exists, $A$ is a bounded linear transformation.*

*Proof.* Let $S_\alpha = \overline{S_\alpha(0)} = \{x \in X / \|x\| \leq \alpha\}$, $D_\alpha = D \cap S_\alpha$. $A$ is defined on each $D_\alpha$. Also $D \supset \bigcup_{n=1}^{\infty} D_n$. If $x \in D$, there exists some positive integer $n$ such that $\|x\| \leq n$ so that $x \in D_n$ and so $x \in \bigcup_{n=1}^{\infty} D_n$. Thus $D \subset \bigcup_{n=1}^{\infty} D_n$ and hence $D = \bigcup_{n=1}^{\infty} D_n$. It follows that

$$A(D) = \bigcup_{n=1}^{\infty} A(D_n).$$

Since $A(D)$ is of the second category, not all of $A(D_n)$ can be nowhere dense. Let $A(D_{n_0})$ be somewhere dense for some positive integer $n_0$. So,

$$(\overline{A(D_{n_0})})^\circ \neq \phi.$$

Thus, there exists some $z \in (\overline{A(D_{n_0})})^\circ$, i.e., $z \in \overline{A(D_{n_0})}$ and there exists a $\delta > 0$ such that

$$B_\delta(z) \subset \overline{A(D_{n_0})}.$$

We now claim that, for $\alpha, \beta, \epsilon > 0$,

$$B_\epsilon(w) \subset \overline{A(D_\alpha)}$$

if and only if

$$B_{\frac{\beta\epsilon}{\alpha}}\left(\frac{\beta w}{\alpha}\right) \subset \overline{A(D_\beta)}.$$

Let $B_\epsilon(w) \subset \overline{A(D_\alpha)}$. Let $z \in B_{\frac{\beta\epsilon}{\alpha}}(\frac{\beta w}{\alpha})$. So,

$$\left\| z - \frac{\beta w}{\alpha} \right\| < \frac{\beta\epsilon}{\alpha},$$

$$i.e., \quad \left\| \frac{\alpha}{\beta} z - w \right\| < \epsilon,$$

$$i.e., \quad \frac{\alpha}{\beta} z \in B_\epsilon(w) \subset \overline{A(D_\alpha)}.$$

Hence, there exists a sequence of points $\{z_k\}$ in $A(D_\alpha)$ such that

$$z_k \to \frac{\alpha}{\beta} z, \ k \to \infty,$$

$$i.e., \quad \frac{\beta}{\alpha} z_k \to z, \ k \to \infty.$$

Since $z_k \in A(D_\alpha)$, we can write

$$z_k = A(x_k), \ \text{where} \ , x_k \in D_\alpha, \ k = 1, 2, \ldots.$$

Also,

$$\left\|\frac{\beta}{\alpha}x_k\right\| \leq \frac{\beta}{\alpha}\alpha = \beta,$$

so that

$$\frac{\beta}{\alpha}x_k \in D_\beta$$

and so

$$\frac{\beta}{\alpha}A(x_k) \in A(D_\beta),$$

$$i.e., \ \frac{\beta}{\alpha}z_k \in A(D_\beta),$$

implying that

$$z \in \overline{A(D_\beta)}.$$

Thus,

$$B_{\frac{\beta\epsilon}{\alpha}}(\frac{\beta w}{\alpha}) \subset \overline{A(D_\beta)}.$$

Consequently,

$$B_\epsilon(w) \subset \overline{A(D_\alpha)} \text{ implies } B_{\frac{\beta\epsilon}{\alpha}}\left(\frac{\beta w}{\alpha}\right) \subset \overline{A(D_\beta)}.$$

The remaining part of the claim can be proved in a similar fashion. Thus the claim is established.

Now,

$$B_{\frac{\delta}{n_0}}\left(\frac{z}{n_0}\right) \subset \overline{A(D_1)}.$$

So far, we have proved the existence of a point $z_1 \in A(D_1)$ such that a neighborhood of this point is wholly contained in $\overline{A(D_1)}$, where we note that such a point $z_1$ exists, because, there must be points $z_1$ in $A(D_1)$, as close to $\frac{z}{n_0}$, as we want, since $\frac{z}{n_0} \in \overline{A(D_1)}$.

$$B_{\delta_1}(z_1) \subset \overline{A(D_1)}.$$

Since $z_1 \in A(D_1)$, there exists $\hat{x} \in D_1$ such that

$$A(\hat{x}) = z_1.$$

Let

$$N = \{x - \hat{x}/x \in D_1\}.$$

For any $x - \hat{x} \in N$, we have

$$\|x - \hat{x}\| \leq \|x\| + \|\hat{x}\| \leq 2, \text{ since } x, \hat{x} \in D_1 \subset S_1.$$

So

$$x - \hat{x} \in S_2 \cap D = D_2, \text{ since } x - \hat{x} \in D,$$

$$i.e., N \subset D_2.$$

Thus,

$$A(N) \subset A(D_2) \text{ and } \overline{A(N)} \subset \overline{A(D_2)}.$$

Now, for any $x - \hat{x} \in N$,

$$A(x - \hat{x}) = A(x) - A(\hat{x}) = A(x) - z_1,$$

which implies that

$$A(N) = A(D_1) - z_1 = \{A(y) - z_1/y \in D_1\},$$

and so

$$\overline{A(N)} = \overline{A(D_1) - z_1} = \overline{A(D_1)} - z_1.$$

We now claim that

$$B_{\frac{\alpha\delta_1}{2}}(0) \subset A(D_\alpha).$$

We note that

$$B_{\delta_1}(0) \subset \overline{A(N)}.$$

For $w \in B_{\delta_1}(0)$, $\|w\| < \delta_1$, i.e.,

$$\|(w + z_1) - z_1\| < \delta_1,$$

$$i.e., \; w + z_1 \in B_{\delta_1}(z_1) \subset \overline{A(D_1)},$$

$$i.e., \; w \in \overline{A(D_1)} - z_1 = \overline{A(N)} \subset \overline{A(D_2)}.$$

Thus,

$$B_{\delta_1}(0) \subset \overline{A(D_2)}$$

and therefore,

$$B_{\frac{\alpha\delta_1}{2}}(0) \subset \overline{A(D_\alpha)},$$

proving our claim. Define

$$P_\alpha = B_{\frac{\alpha\delta_1}{2}}(0), \text{ where } \alpha = \frac{1}{2^k}, \; k = 0, 1, 2, \ldots.$$

Then

$$P_1 \subset A(D_2) \subset A(D).$$

We note that for any $k = 0, 1, 2, \ldots,$

$$P_{2^{-k}} \subset \overline{A(D_{2^{-k}})}$$

and the radius of $P_{2^{-k}}$ is $\frac{\delta_1}{2^{k+1}}$. If $y \in P_1$, then $y \in \overline{A(D_1)}$. So, there exists some $v \in A(D_1)$ such that $\|y - v\| < \frac{\delta_1}{2^2}$. Since $v \in A(D_1)$, $v = A(x_1)$, for some $x_1 \in D$, i.e., there exists $x_1 \in D$ such that

$$\|y - A(x_1)\| < \frac{\delta_1}{2^2},$$

from which we get

$$y - A(x_1) \in P_{2^{-1}} \subset \overline{A(D_{2^{-1}})}.$$

Arguing as above, we assert the existence of $x_2 \in D_{2^{-1}}$ such that

$$\|y - A(x_1) - A(x_2)\| < \frac{\delta_1}{2^3}.$$

Proceeding like this, we can construct the sequence $\{x_k\}$ such that $x_k \in D_{2^{-k+1}}$ and

$$\|y - A(x_1) - A(x_2) - \cdots - A(x_n)\| < \frac{\delta_1}{2^{n+1}}.$$

Let $s_n = \sum_{k=1}^{n} x_k$, $n = 1, 2, \ldots$. Then $\|s_n\| \le \sum_{k=1}^{n} \|x_k\|$. Since $x_k \in D_{2^{-k+1}}$, $k = 0, 1, 2, \ldots$, $\|x_1\| \le 1$, $\|x_2\| \le \frac{1}{2}$, $\|x_3\| \le \frac{1}{2^2}$, $\ldots$, so that $\|s_n\| \le 2$. Thus,

$s_n \in S_2$, $n = 1, 2, \ldots$. Also, $s_n \in D$, $n = 1, 2, \ldots$. $\{s_n\}$ is clearly a Cauchy sequence in $X$ and since $X$ is a Banach space, there exists $x \in X$ such that $s_n \to x$, $n \to \infty$. $\|x\| \leq 2$, since $\|s_n\| \leq 2$, for all $n = 1, 2, \ldots$ and so $x \in S_2$. Since

$$\|y - A(s_n)\| < \frac{\delta_1}{2^{n+1}}, \; A(s_n) \to y, \; n \to \infty.$$

Since $A$ is a closed linear transformation, $x \in D$ and $y = A(x)$; since $x \in S_2$, $x \in D_2$. Since $y \in P_1$, $P_1 \subset A(D)$. We shall now prove that $A(D) = Y$. Let $y \in Y$ and $y \neq 0$. Let $\beta = \frac{\delta_1}{4\|y\|}$. Then, $\beta y \in P_1$. Since $P_1 \subset A(D)$, there exists $x \in D$ such that $\beta y = A(x)$, i.e., $y = A\left(\frac{x}{\beta}\right)$, implying that $y \in A(D)$. Note that the result holds trivially if $y = 0$. Thus $A(D) = Y$, proving that $A$ is onto, i.e., (i) holds.

To prove (ii), let $y \in Y$. We can find a scalar $\beta$ and some $x_0 \in D_2$ such that $y = A\left(\frac{x_0}{\beta}\right)$. Clearly,

$$\left\|\frac{x_0}{\beta}\right\| = \frac{\|x_0\|}{\beta} \leq \frac{2}{\beta} = \frac{8\|y\|}{\delta_1}, \text{ where } \beta \text{ was chosen above.}$$

Taking $x = \frac{x_0}{\beta}$, $m = \frac{8}{\delta_1}$, (ii) follows.

Finally, we claim that (ii) implies (iii). Let $y \in Y$ and $A^{-1}$ exist. Then there exists $x \in X$ such that

$$x = A^{-1}(y).$$

We now have

$$\|x\| = \|A^{-1}(y)\| \leq m\|y\|, \text{ for all } y \in Y,$$

proving that $A^{-1}$ is a bounded linear transformation. The proof of the theorem is now complete. □

## 6.2 Some consequences of the bounded inverse theorem: Closed graph theorem

In this section, the main theorem that we prove is the "closed graph theorem" using the bounded inverse theorem.

**Theorem 60.** *Let $X, Y$ be Banach spaces and $A : X \to Y$ be a bounded linear transformation of $X$ onto $Y$. If $A^{-1}$ exists, then $A^{-1}$ is a bounded linear transformation.*

*Proof.* Since $A$ is defined on the whole of $X$ and $A$ is bounded, $A$ is closed. Since $A$ is onto and $Y$ is a Banach space, $A(X)$ is of the second category in view of Baire's theorem. Now, using the bounded inverse theorem, the present result follows. □

**Theorem 61** (Closed graph theorem)**.** *Let $X$ and $Y$ be Banach spaces and $A$ be a linear transformation of $X$ into $Y$. If $A$ is closed, then $A$ is bounded.*

*Proof.* Since $X$ and $Y$ are Banach spaces, $X \times Y$ is also a Banach space under the norm

$$\|(x, y)\| = \|x\| + \|y\|, \ (x, y) \in X \times Y.$$

Since $A$ is closed, $G_A$ is a closed subspace of $X \times Y$, which is complete. Hence $G_A$ is complete, i.e., $G_A$ is a Banach space. Now, consider the mapping

$$B : G_A \to X$$

defined by

$$(x, A(x)) \to x.$$

Clearly $B$ is linear. Since

$$\begin{aligned} \|B(x, A(x))\| = \|x\| &\leq \|x\| + \|A(x)\| \\ &= \|(x, A(x))\|, \end{aligned}$$

$B$ is bounded. The reader can check that $B$ is surjective, i.e., onto. $B$ is also injective, i.e., one-to-one, since, if

$$
\begin{aligned}
&B(x, A(x)) = 0,\\
\textit{i.e.,}\quad & x = 0,\\
\textit{i.e.,}\quad & A(x) = 0,\\
\textit{i.e.,}\quad & (x, A(x)) = (0, 0).
\end{aligned}
$$

Consequently, $B^{-1}$ exists. Hence, we have the following:

(i) $G_A$ and $X$ are Banach spaces;

(ii) $B$ is a bounded linear transformation, which is also onto;

and

(iii) $B^{-1}$ exists.

Using Theorem 60, $B^{-1}$ is a bounded linear transformation. Let $x_n \to x$, $n \to \infty$ in $X$. Since $B^{-1}$ is continuous,

$$
\begin{aligned}
&B^{-1}(x_n) \to B^{-1}(x),\ n \to \infty,\\
\textit{i.e.,}\quad & (x_n, A(x_n)) \to (x, A(x)),\ n \to \infty,\\
\textit{i.e.,}\quad & (x_n - x, A(x_n) - A(x)) \to (0, 0),\ n \to \infty,
\end{aligned}
$$

which implies that

$$A(x_n) \to A(x),\ n \to \infty.$$

Thus $A$ is continuous. This completes the proof of the closed graph theorem. □

We now consider an application of the closed graph theorem.

**Definition 62.** *A linear transformation $P$ on a linear space $X$ is called a "projection" if $P^2 = P$.*

**Definition 63.** *Let $M, N$ be subspaces of a linear space $X$. If every $z \in X$ can be expressed as $z = x + y$, where $x \in M$, $y \in N$, we say that $X$ is the "direct sum" of $M$ and $N$, written as*

$$X = M \oplus N.$$

The following results are easily proved.

**Theorem 62.** *Let $P$ be a projection on a Banach space $X$. Then*

*(i)* $I - P$ *is a projection on* $X$*;*

*(ii)* $\Re(P) = \{x \in X/P(x) = x\}$, *where* $\Re(P)$ *denotes the range of* $P$*;*

*(iii)* $\Re(P) = \aleph(I - P)$, *where* $\aleph(T) = \{x \in X/T(x) = 0\}$*;*

*(iv)* $X = \Re(P) \oplus \Re(I - P)$*;*

*and*

*(v) if* $P$ *is bounded, then* $\Re(P)$ *and* $\Re(I - P)$ *are closed.*

**Theorem 63.** *Let $M, N$ be linear subspaces of $X$ with $X = M \oplus N$. Then there exists a unique projection $P$ on $X$ with $\Re(P) = M$ and $\Re(I - P) = N$.*

Note that if $P$ is bounded, $M$ and $N$ are closed subspaces of $X$.

The following result shows that every direct sum decomposition of $X$, formed from closed subspaces of $X$, arises from a bounded projection on $X$.

**Theorem 64.** *Let $X$ be a Banach space. Let $M, N$ be closed, linear subspaces of $X$ such that $X = M \oplus N$. Then there exists a unique bounded projection $P$ on $X$ such that $\Re(P) = M$ and $\Re(I - P) = N$.*

*Proof.* It is clear that there exists a unique projection $P$ on $X$ such that $\Re(P) = M$ and $\Re(I - P) = N$. We claim that $P$ is closed. Let $\{x_n\}$ be a sequence in $X$ such that $x_n \to x$, $n \to \infty$ and $P(x_n) \to y$, $n \to \infty$. Since $P(x_n) \in M$, $n = 1, 2, \ldots$ and $M$ is closed, $y \in M$. We also have,

$$x_n - P(x_n) = (I - P)(x_n) \in N.$$

Since $N$ is closed, $x - y = \lim_{n \to \infty} (x_n - P(x_n)) \in N$ and so

$$\begin{aligned} & P(x-y) = 0, \\ i.e., \quad & P(x) - P(y) = 0, \\ i.e., \quad & P(x) - y = 0, \text{ since } y \in M,\ P(y) = y \\ i.e., \quad & y = P(x). \end{aligned}$$

Thus $P$ is closed. In view of the closed graph theorem, $P$ is bounded, completing the proof of the theorem. □

---

## 6.3 Interior mapping principle and the open mapping theorem

**Theorem 65** (Interior mapping principle). *With the same assumptions of the "Bounded inverse theorem," let $E$ be a relatively open subset of $D$, i.e., $E = D \cap O$, where $O$ is open in $X$. Then $A(E)$ is open in $Y$.*

*Proof.* We first claim that for each $\alpha > 0$, there corresponds some $\beta > 0$ such that

$$B_\beta \subset A(D \cap S_\alpha),$$

where

$$S_\alpha = \{x \in X / \|x\| \leq \alpha\} \text{ and } B_\beta = \{y \in Y / \|y\| \leq \beta\}.$$

This is clear from the "bounded inverse theorem." In fact, it was proved that

$$\begin{aligned} & P_1 \subset A(D_2), \\ i.e., \quad & B_{\frac{\delta_1}{2}} \subset A(D_2), \end{aligned}$$

which implies that

$$B_{\frac{\alpha\delta_1}{4}} \subset A(D_\alpha).$$

Taking $\beta = \frac{\alpha\delta_1}{4}$, our claim is proved. We now prove the theorem. Let $E = D \cap O$, where $O$ is an open set in $X$. Let $y_0 \in A(E)$, so that for some $x_0 \in E$, $y_0 = A(x_0)$. The set $-x_0 + O$ is a neighborhood of the origin and so contains $S_\alpha$, for some $\alpha > 0$.

Then

$$D \cap S_\alpha \subset D \cap (-x_0 + O) = -x_0 + D \cap O.$$

Using the claim proved above, there exists $\beta > 0$ such that

$$B_\beta \subset A(D \cap S_\alpha) \subset A(-x_0 + D \cap O) = -y_0 + A(E),$$

which implies that

$$B_\beta + y_0 \subset A(E).$$

Since $y_0 + B_\beta$ contains a neighborhood of $y_0$, $A(E)$ contains a neighborhood of $y_0$. Since $y_0 \in A(E)$ is arbitrary, it follows that $A(E)$ is open in $Y$, completing the proof. □

The "open mapping theorem" now follows:

**Theorem 66** (Open mapping theorem). *If $X, Y$ are Banach spaces and $A$ is a continuous linear transformation of $X$ onto $Y$, then $A$ is an open mapping.*

*Proof.* This follows immediately from Theorem 65, i.e., the "interior mapping principle," by taking $X$ in the place of $D$. □

**Corollary 5.** *A one-to-one continuous linear mapping of a Banach space $X$ onto a Banach space $Y$, over the same field, is a linear homeomorphism of $X$ onto $Y$.*

*Proof.* Let $A$ be a continuous one-to-one linear mapping of $X$ onto $Y$. Then, by the open mapping theorem, $A(U)$ is an open set of $Y$ whenever $U$ is an open set in $X$. Since $A$ is one-to-one, $(A^{-1})^{-1}(U) = A(U)$, i.e., $(A^{-1})^{-1}(U)$ is open whenever $U$ is open, i.e., $A^{-1}$ is continuous. Thus $A$ is a linear homeomorphism of $X$ onto $Y$, completing the proof. □

**Exercises**

1. Prove Theorem 62.

2. Prove Theorem 63.

3. If $X, Y$ are Banach spaces, prove that $X \times Y$ is also a Banach space under the norm

$$\|(x, y)\| = \|x\| + \|y\|, \ (x, y) \in X \times Y.$$

4. Repeat Exercise 3 under the norm

$$\|(x, y)\| = \max(\|x\|, \|y\|), \ (x, y) \in X \times Y.$$

5. Let $X, Y$ be normed linear spaces and $A : X \to Y$ be closed. If $F$ is a compact subset of $X$, prove that $A(F)$ is a closed subset of $Y$.

6. Let $X$ be a Banach space under the two norms $\|.\|_1$ and $\|.\|_2$. If there exists $\alpha > 0$ such that $\|x\|_1 \leq \alpha\|x\|_2$, for every $x \in X$, prove that $\|.\|_1$ and $\|.\|_2$ are equivalent.

7. Let $X$ be a Banach space under the two norms $\|.\|_1$ and $\|.\|_2$. If $\|x_n\|_1 \to 0$, $n \to \infty$ implies that $\|x_n\|_2 \to 0$, $n \to \infty$, prove that $\|.\|_1$ and $\|.\|_2$ are equivalent.

8. Can the completeness assumption in the closed graph theorem be omitted? Justify your contention.

9. Can the completeness assumptions on $X$ and $Y$ in the open mapping theorem be dropped? Justify your answer.

10. Let $X, Y$ be Banach spaces and $F$ be a bounded linear transformation of $X$ into $Y$. Prove that $\Re(F)$ is closed in $Y$ if and only if $\Re(F)$ is linearly homeomorphic to $\frac{X}{\aleph(F)}$.

11. Let $X, Y$ be Banach spaces and $F : X \to Y$ be a bounded, one-to-one, linear mapping. Prove that the range of $F$, i.e., $\Re(F)$, is closed in $Y$ if and only if $F^{-1} : \Re(F) \to X$ is bounded.

12. Prove that even a discontinuous linear mapping can have a closed graph. Does this contradict the closed graph theorem?

# Chapter 7

# Hilbert Spaces

## 7.1 Hermitian forms, inner product spaces, and Hilbert spaces

In the sequel, we suppose that $X$ is a linear space over the field $\mathbb{C}$ of complex numbers, unless otherwise stated.

**Definition 64.** *By a sesquilinear Hermitian form on $X$, we mean a mapping*

$$\varphi : X \times X \to \mathbb{C}$$

*satisfying the following conditions:*

*(i)* $\varphi(x_1 + x_2, y) = \varphi(x_1, y) + \varphi(x_2, y)$,

$\varphi(x, y_1 + y_2) = \varphi(x, y_1) + \varphi(x, y_2)$, $x, y, x_1, x_2, y_1, y_2 \in X$;

*(ii)* $\varphi(\lambda x, y) = \lambda\varphi(x, y)$,

$\varphi(x, \lambda y) = \overline{\lambda}\varphi(x, y)$, $x, y \in X$, $\lambda \in \mathbb{C}$;

*and*

*(iii)* $\varphi(x, y) = \overline{\varphi(y, x)}$, $x, y \in X$.

*From the above definition, it follows that*

$$\varphi\left(\sum_{j=1}^{n} \lambda_j x_j, \sum_{k=1}^{m} \mu_k y_k\right) = \sum_{j=1}^{n}\sum_{k=1}^{m} \lambda_j \overline{\mu_k} \varphi(x_j, y_k).$$

**Definition 65.** *Given a sesquilinear Hermitian form $\varphi$ on $X$, define*

$$\tilde{\varphi} : X \to \mathbb{C}$$

*by*

$$\tilde{\varphi}(x) = \varphi(x, x), \text{ for all } x \in X.$$

*In view of Definition 64 (iii), $\tilde{\varphi}(x)$ is always real. This induced mapping $\tilde{\varphi}$ from $\varphi$ is called a "quadratic form" on $X$.*

The following is a very useful formula, the proof of which is straightforward and is left to the reader.

**Theorem 67** (Polarization formula)**.** *If $\tilde{\varphi}$ is a quadratic form induced by a Hermitian form $\varphi$ on $X$, then*

$$4\varphi(x, y) = \tilde{\varphi}(x + y) - \tilde{\varphi}(x - y) + i\{\tilde{\varphi}(x + iy) - \tilde{\varphi}(x - iy)\}.$$

**Definition 66.** *A Hermitian form $\varphi$ on $X$ is said to be "degenerate" if there exists $x \in X$, $x \neq 0$ such that*

$$\varphi(x, y) = 0, \text{ for all } y \in X$$

*and $\varphi$ is said to be "non-degenerate" otherwise, i.e., for any $x \in X$, $x \neq 0$, there exists $y \in X$ such that $\varphi(x, y) \neq 0$.*

**Definition 67.** *A Hermitian form $\varphi$ on $X$ is said to be "positive," if $\varphi(x, x) \geq 0$, for all $x \in X$.*

**Theorem 68** (Schwarz inequality)**.** *If $\varphi$ is a positive Hermitian form on $X$, then*

$$|\varphi(x, y)|^2 \leq \varphi(x, x)\varphi(y, y), \text{ for all } x, y \in X.$$

*Proof.* Suppose $\varphi(y, y) \neq 0$. If $\lambda \in \mathbb{C}$, then

$$\varphi(x + \lambda y, x + \lambda y) \geq 0$$

and

$$\varphi(y, y)\varphi(x + \lambda y, x + \lambda y) \geq 0.$$

So

$$\begin{aligned}
&\varphi(y,y)[\varphi(x,x)+\varphi(x,\lambda y)+\varphi(\lambda y,x)+\varphi(\lambda y,\lambda y)]\geq 0,\\
\textit{i.e.,}\quad &\varphi(y,y)[\varphi(x,x)+\overline{\lambda}\varphi(x,y)+\lambda\varphi(y,x)+\lambda\overline{\lambda}\varphi(y,y)]\geq 0,\\
\textit{i.e.,}\quad &\varphi(y,y)[\varphi(x,x)+\overline{\lambda}\varphi(x,y)+\lambda\varphi(y,x)+|\lambda|^2\varphi(y,y)]\geq 0,\\
\textit{i.e.,}\quad &\varphi(x,x)\varphi(y,y)+\lambda\varphi(y,y)\varphi(y,x)+\overline{\lambda}\varphi(y,y)\varphi(x,y)+|\lambda|^2(\varphi(y,y))^2\geq 0,\\
\textit{i.e.,}\quad &\varphi(x,x)\varphi(y,y)-|\varphi(x,y)|^2+\varphi(x,y)\overline{\varphi(x,y)}+\lambda\varphi(y,y)\varphi(y,x)\\
&\qquad +\overline{\lambda}\varphi(y,y)\varphi(x,y)+|\lambda|^2(\varphi(y,y))^2\geq 0,\\
\textit{i.e.,}\quad &\varphi(x,x)\varphi(y,y)-|\varphi(x,y)|^2+(\overline{\lambda}\varphi(y,y)+\overline{\varphi(x,y)})\varphi(x,y)\\
&\qquad +(\varphi(y,x)+\overline{\lambda}\varphi(y,y))\lambda\varphi(y,y)\geq 0,\\
\textit{i.e.,}\quad &\varphi(x,x)\varphi(y,y)-|\varphi(x,y)|^2+\{\overline{\lambda}\varphi(y,y)+\overline{\varphi(x,y)}\}(\varphi(x,y)+\lambda\varphi(y,y))\geq 0,\\
\textit{i.e.,}\quad &\varphi(x,x)\varphi(y,y)-|\varphi(x,y)|^2+\{\varphi(x,y)+\lambda\varphi(y,y)\}\{\overline{\varphi(x,y)+\lambda\varphi(y,y)}\}\geq 0,\\
\textit{i.e.,}\quad &\varphi(x,x)\varphi(y,y)-|\varphi(x,y)|^2+2RP\{\varphi(x,y)+\lambda\varphi(y,y)\}\geq 0.
\end{aligned}$$

Choose $\lambda=-\frac{\varphi(x,y)}{\varphi(y,y)}$. Then we have

$$\begin{aligned}
&\varphi(x,x)\varphi(y,y)-|\varphi(x,y)|^2+2RP\left\{\varphi(x,y)-\frac{\varphi(x,y)}{\varphi(y,y)}\varphi(y,y)\right\}\geq 0,\\
\textit{i.e.,}\quad &\varphi(x,x)\varphi(y,y)-|\varphi(x,y)|^2\geq 0,\\
\textit{i.e.,}\quad &|\varphi(x,y)|^2\leq\varphi(x,x)\varphi(y,y).
\end{aligned}$$

A similar argument holds if $\varphi(x,x)\neq 0$. If $\varphi(x,x)=\varphi(y,y)=0$,

$$\varphi(x+\lambda y,x+\lambda y)\geq 0,\ \text{for all } \lambda\in\mathbb{C}.$$

Then

$$\begin{aligned}
&\varphi(x,x)+\overline{\lambda}\varphi(x,y)+\lambda\varphi(y,x)+\lambda\overline{\lambda}\varphi(y,y)\geq 0,\\
\textit{i.e.,}\quad &\overline{\lambda}\varphi(x,y)+\lambda\varphi(y,x)\geq 0,\ \text{since } \varphi(x,x)=\varphi(y,y)=0.
\end{aligned}$$

Choose $\lambda=-\varphi(x,y)$. Then

$$\begin{aligned}
&-\overline{\varphi(x,y)}\varphi(x,y)-\varphi(x,y)\overline{\varphi(x,y)}\geq 0,\\
\textit{i.e.,}\quad &-2|\varphi(x,y)|^2\geq 0,
\end{aligned}$$

which implies that $\varphi(x,y)=0$. Consequently, the inequality holds trivially in this case, completing the proof. □

**Corollary 6.** *If $\varphi$ is a positive Hermitian form on $X$, the set of all $x \in X$ such that $\varphi(x,x) = 0$ is precisely the subspace of $X$ consisting of all $x \in X$ such that $\varphi(x,y) = 0$, for all $y \in X$.*

**Corollary 7.** *A positive Hermitian form $\varphi$ on $X$ is non-degenerate if and only if*

$$\varphi(x,x) > 0, \text{ for all } x \in X, x \neq 0.$$

If $\varphi$ is a non-degenerate, positive Hermitian form on $X$, then

$$\|x\| = +\sqrt{\varphi(x,x)}, \; x \in X$$

is a norm on $X$. It suffices to prove the triangle inequality. If $x, y \in X$,

$$\begin{aligned}
\varphi(x+y, x+y) &= \varphi(x,x) + \varphi(x,y) + \varphi(y,x) + \varphi(y,y) \\
&= \varphi(x,x) + \{\varphi(x,y) + \overline{\varphi(x,y)}\} + \varphi(y,y) \\
&= \varphi(x,x) + 2RP\varphi(x,y) + \varphi(y,y) \\
&\leq \varphi(x,x) + 2|\varphi(x,y)| + \varphi(y,y) \\
&\leq \varphi(x,x) + 2\{\varphi(x,x)\}^{\frac{1}{2}}\{\varphi(y,y)\}^{\frac{1}{2}} + \varphi(y,y) \\
&= [\{\varphi(x,x)\}^{\frac{1}{2}} + \{\varphi(y,y)\}^{\frac{1}{2}}]^2,
\end{aligned}$$

using Schwarz inequality.

Thus,

$$\{\varphi(x+y, x+y)\}^{\frac{1}{2}} \leq \{\varphi(x,x)\}^{\frac{1}{2}} + \{\varphi(y,y)\}^{\frac{1}{2}},$$

$$i.e., \; \|x+y\| \leq \|x\| + \|y\|, \; x, y \in X,$$

completing the proof.

**Corollary 8.** *Let $\varphi$ be a positive Hermitian form. The proof of the Schwarz inequality shows that*

$$|\varphi(x,y)|^2 = \varphi(x,x)\varphi(y,y)$$

*if and only if there exists $\lambda \in \mathbb{C}$ such that*

$$\varphi(x + \lambda y, x + \lambda y) = 0.$$

*Now, suppose $\varphi$ is non-degenerate. Then the above implies that $x + \lambda y = 0$, i.e., $x, y$ are linearly dependent. Note that the converse is also true.*

**Definition 68.** *A linear space $X$ over $\mathbb{C}$ is called an "inner product space" if $X$ is provided with a non-degenerate, positive, Hermitian form $\varphi$. $X$ is called a "Hilbert space" if $X$ is an inner product space such that it is complete with respect to the metric defined by the norm induced by $\varphi$, i.e.,*

$$d(x, y) = \|x - y\| = +\sqrt{\varphi(x - y, x - y)}, \text{ for } x, y \in X.$$

The value $\varphi(x, y) \in \mathbb{C}$ is called the "inner product" or "scalar product" of $x$ and $y$. Hereafter, we write $(x, y)$ in the place of $\varphi(x, y)$ and as before, $\|x\|$ in the place of $\sqrt{\varphi(x, x)}$ so that $\|x\|^2 = (x, x)$, if $X$ is an inner product space and $x, y \in X$. For convenience, we denote an inner product by $(x, y)$ in this chapter (instead of $< x, y >$). In the context of inner product spaces, there is the concept of "orthogonality," which enables us to carry over the geometry in the Euclidean space $\mathbb{C}^n$ or $\mathbb{R}^n$ to these spaces.

**Definition 69.** *If $x, y \in X$, $x$ is said to be "orthogonal" to $y$, read as $x$ is perpendicular to $y$, written as $x \perp y$, if*

$$(x, y) = 0.$$

Note that the Schwarz inequality holds in an inner product space, i.e.,

$$|(x, y)| \leq \|x\|\|y\|, \text{ for all } x, y \in X.$$

**Theorem 69.** *(i) for $x, y \in X$,*

$$\|x + y\|^2 + \|x - y\|^2 = 2\|x\|^2 + 2\|y\|^2 \text{ (parallelogram law);}$$

*(ii) if $x \perp y$, then*

$$\|x + y\|^2 = \|x\|^2 + \|y\|^2 \text{ (Pythagorean theorem).}$$

*Proof.* (i)

$$\begin{aligned}\|x+y\|^2+\|x-y\|^2 &= (x+y,x+y)+(x-y,x-y)\\ &= (x,x)+(x,y)+(y,x)+(y,y)\\ &\quad +(x,x)-(x,y)-(y,x)+(y,y)\\ &= 2(x,x)+2(y,y)\\ &= 2\|x\|^2+2\|y\|^2;\end{aligned}$$

(ii) if $x \perp y$, then $(x,y)=0$, so that

$$\begin{aligned}\|x+y\|^2 &= (x+y,x+y)\\ &= (x,x)+(x,y)+(y,x)+(y,y)\\ &= \|x\|^2+\|y\|^2, \text{ since } (x,y)=0 \text{ and}\\ &\qquad (y,x)=\overline{(x,y)}=\bar{0}=0.\end{aligned}$$

Hence the result. □

**Definition 70.** *A family $(x_a)_{a\in\Lambda}$ is said to be orthogonal, if $x_a \perp x_b$ whenever $a,b\in\Lambda$, $a\neq b$.*

It is easy to prove that if $\{x_1,x_2,\ldots,x_n\}$ is a finite orthogonal family, then

$$\left\|\sum_{j=1}^{n} x_j\right\|^2 = \sum_{j=1}^{n}\|x_j\|^2.$$

**Definition 71.** *An element $x\in X$ is said to be a "normalized vector" if $\|x\|=1$.*

**Definition 72.** *If $x\in X$, $x\neq 0$, the mapping*

$$x \to \frac{x}{\|x\|}$$

*is called a "normalization."*

**Definition 73.** *A family $(x_a)_{a\in\Lambda}$ in an inner product space $X$ is said to be an "orthonormal family," if it is an orthogonal family in which every element is normalized.*

**Theorem 70** (Bessel's inequality). *If $\{x_j\}_{j=1}^n$ is a finite orthonormal family in an inner product space $X$, then*

$$\sum_{j=1}^{n} |(x, x_j)|^2 \leq \|x\|^2, \text{ for all } x \in X.$$

*Proof.* Consider the element $y = x - \sum_{j=1}^{n}(x, x_j)x_j$ in $X$. Then $\|y\| \geq 0$ and

$$\begin{aligned}
\|y\|^2 &= \left\| x - \sum_{j=1}^{n}(x, x_j)x_j \right\|^2 \\
&= \left( x - \sum_{j=1}^{n}(x, x_j)x_j, x - \sum_{j=1}^{n}(x, x_j)x_j \right) \\
&= \|x\|^2 - \sum_{j=1}^{n} \overline{(x, x_j)}(x, x_j) - \sum_{j=1}^{n}(x, x_j)(x_j, x) + \sum_{j,k=1}^{n} (x, x_j)\overline{(x, x_k)}(x_j, x_k) \\
&= \|x\|^2 - 2\sum_{j=1}^{n} |(x, x_j)|^2 + \sum_{j=1}^{n} |(x, x_j)|^2, \text{ since} \\
&\qquad (x_j, x_k) = 1 \text{ if } k = j \text{ and } (x_j, x_k) = 0, \text{ if } k \neq j \\
&= \|x\|^2 - \sum_{j=1}^{n} |(x, x_j)|^2,
\end{aligned}$$

from which we have

$$\sum_{j=1}^{n} |(x, x_j)|^2 \leq \|x\|^2, \text{ for all } x \in X.$$

□

We can deduce Schwarz inequality from Bessel's inequality, which is left as an exercise for the reader.

## 7.2 Orthogonality and orthogonal complements

**Definition 74.** *Let $M$ be a subset of an inner product space $X$. An element $x \in X$ is said to be "orthogonal to $M$" if*

$$(x, y) = 0, \text{ for every } y \in M,$$

*which we write as $x \perp M$ and read as "$x$ is perpendicular to $M$."*

**Definition 75.** *Let $M_1, M_2$ be subsets of the inner product space $X$. $M_1, M_2$ are said to be orthogonal to each other if*

$$(x, y) = 0, \text{ for every } x \in M_1 \text{ and } y \in M_2,$$

*which we write as $M_1 \perp M_2$ and read as "$M_1$ is perpendicular to $M_2$" (or vice versa).*

**Definition 76.** *Let $H$ be a Hilbert space and $M$ be a subset of $H$. Then, the set*

$$\{x \in H / x \perp M\}$$

*is denoted by $M^\perp$, which is called the "orthogonal complement" of $M$.*

We now prove an important proposition.

**Theorem 71.** *We have the following:*

*(i)* $M^\perp$ *is a closed subspace of* $H$ *and*

$M \cap M^\perp = \phi$ *or* $M \cap M^\perp = \{0\}$;

*(ii)* $M \subset (M^\perp)^\perp$;

*(iii) if $N$ is a subset of $H$ such that $M \subset N$, then*

$$N^\perp \subset M^\perp;$$

*and*

*(iv)* $M^{\perp} = ((M^{\perp})^{\perp})^{\perp}$.

*Proof.* (i) Let $x \in M$, $y_1, y_2 \in M^{\perp}$, $\lambda_1, \lambda_2 \in \mathbb{C}$. Then

$$\begin{aligned}(x, \lambda_1 y_1 + \lambda_2 y_2) &= \overline{\lambda_1}(x, y_1) + \overline{\lambda_2}(x, y_2) \\ &= 0,\end{aligned}$$

and so $\lambda_1 y_1 + \lambda_2 y_2 \in M^{\perp}$, proving that $M^{\perp}$ is a subspace of $X$. We now claim that $M^{\perp}$ is closed. Let $\{y_n\}$ be a sequence in $M^{\perp}$ such that

$$y_n \to y, \ n \to \infty, \ y \in X.$$

So, given $\epsilon > 0$, there exists a positive integer $n_0$ such that

$$\|y_n - y\| < \frac{\epsilon}{\|x\|},$$

for a given $x \in M$. Now,

$$\begin{aligned}|(x, y)| &= |(x, y) - (x, y_n) + (x, y_n)| \\ &\leq |(x, y) - (x, y_n)| + |(x, y_n)| \\ &= |(x, y - y_n)|, \text{ since } (x, y_n) = 0, \ n = 1, 2, \ldots \\ &\leq \|x\| \|y - y_n\| \\ &< \|x\| \frac{\epsilon}{\|x\|}, \text{ for all } n \geq n_0 \\ &= \epsilon.\end{aligned}$$

Since $\epsilon > 0$ is arbitrary, $(x, y) = 0$, which implies that $y \in M^{\perp}$, proving our claim. If $M \cap M^{\perp} = \phi$, we are through. If $M \cap M^{\perp} \neq \phi$, then $x \in M \cap M^{\perp}$, for some $x \in X$. Suppose $x \neq 0$. Then $x \perp x$, i.e., $(x, x) = 0$, i.e., $\|x\|^2 = 0$ and so $x = 0$, which is a contradiction. Hence $x = 0$. Thus $M \cap M^{\perp} = \{0\}$. This proves (i).

(ii) If $x \in M$, $y \in M^{\perp}$, then $x \perp y$ so that $x \perp M^{\perp}$ and consequently $x \in (M^{\perp})^{\perp}$, i.e., $M \subset (M^{\perp})^{\perp}$, proving (ii).

(iii) If $y \in N^{\perp}$, $y \perp N$. Since $M \subset N$, $y \perp M$ and so $y \in M^{\perp}$. Thus, $N^{\perp} \subset M^{\perp}$.

(iv) From (ii), $M^\perp \subset ((M^\perp)^\perp)^\perp$ and from the fact that $M \subset (M^\perp)^\perp$, it follows that $((M^\perp)^\perp)^\perp \subset M^\perp$, using (iii) and consequently,

$$M^\perp = ((M^\perp)^\perp)^\perp.$$

The proof of the theorem is now complete. □

**Theorem 72.** *Let $M$ be a closed subspace of a Hilbert space $H$ and $x \in H$. If*

$$d = \inf_{y \in M} \{\|y - x\|\},$$

*then there exists a unique element $y_0 \in M$ such that $\|y_0 - x\| = d$.*

*Proof.* Given $\epsilon > 0$, there exists $y \in M$ such that

$$d \leq \|y - x\| < d + \epsilon, \text{ using the definition of infimum.}$$

Choose $\epsilon = \frac{1}{n}$. Then there exists $y_n \in M$ such that

$$d \leq \|y_n - x\| < d + \frac{1}{n}, \; n = 1, 2, \ldots.$$

Thus, $\{y_n\}$ is a sequence in $M$ such that

$$\|y_n - x\| \to d, \; n \to \infty.$$

Using the parallelogram law we have

$$\begin{aligned}\|y_n - y_m\|^2 &= \|(y_n - x) - (y_m - x)\|^2 \\ &= 2\|y_n - x\|^2 + 2\|y_m - x\|^2 - 4\|\frac{1}{2}(y_n + y_m) - x\|^2, \text{ for all } n, m\end{aligned}$$

(the reader can work out the details).

Note that $\frac{1}{2}(y_n + y_m) \in M$, since $M$ is a subspace of $H$ and so

$$\|\frac{1}{2}(y_n + y_m) - x\|^2 \geq d^2.$$

Hence,

$$\begin{aligned}\|y_n - y_m\|^2 &\leq 2\|y_n - x\|^2 + 2\|y_m - x\|^2 - 4d^2 \\ &\to 2d^2 + 2d^2 - 4d^2 = 0, \; n, m \to \infty.\end{aligned}$$

Consequently, $\{y_n\}$ is a Cauchy sequence in $M$. Since $M$ is a closed subspace of $H$, $M$ is complete and so

$$y_n \to y_0, \ n \to \infty \text{ in } M.$$

Since $\|y_n - x\| \to d$, $n \to \infty$ and $y_n \to y_0$, $n \to \infty$, it follows that

$$\|y_0 - x\| = d.$$

We claim that $y_0$ is unique. Suppose $z_0 \in M$ is such that

$$\|z_0 - x\| = d.$$

Using the parallelogram law,

$$\begin{aligned}
&\|(y_0 - x) + (z_0 - x)\|^2 + \|(y_0 - x) - (z_0 - x)\|^2 \\
&\quad = 2\|y_0 - x\|^2 + 2\|z_0 - x\|^2, \\
\text{i.e.,} \quad &\|y_0 + z_0 - 2x\|^2 + \|y_0 - z_0\|^2 = 2d^2 + 2d^2 = 4d^2, \\
\text{i.e.,} \quad &\|y_0 - z_0\|^2 = 4d^2 - \|y_0 + z_0 - 2x\|^2 \\
&\qquad = 4d^2 - 4\left\|\frac{y_0 + z_0}{2} - x\right\|^2.
\end{aligned}$$

Since $M$ is a subspace of $H$, $\frac{y_0+z_0}{2} \in M$ and so

$$\left\|\frac{y_0 + z_0}{2} - x\right\|^2 \geq d^2,$$

so that

$$\|y_0 - z_0\|^2 \leq 4d^2 - 4d^2 = 0,$$

from which it follows that $\|y_0 - z_0\| = 0$, i.e., $y_0 = z_0$. Thus $y_0$ is unique, completing the proof of the theorem. □

**Note 13.** *For obvious reasons, it suffices to assume that $M$ is convex, which is complete under the induced metric.*

**Theorem 73.** *Let $M$ be a closed subspace of a Hilbert space $H$ and $N$ be a proper closed subspace of $M$. Then there exists $x \in M$, $x \neq 0$ such that $x \perp N$.*

*Proof.* Let $z \in M$, $z \notin N$. Let

$$d = \inf_{y \in N} \|y - z\|.$$

Then, in view of Theorem 72, there exists $y_0 \in N$ such that

$$d = \|y_0 - z\|.$$

Since $y_0 \in N$ and $z \notin N$, $z - y_0 \neq 0$. We shall now prove that $x = y_0 - z \perp N$. Let $y \in N$. Then $y_0 + \lambda y \in N$, for any $\lambda \in \mathbb{C}$.

$$\|x + \lambda y\| = \|y_0 + \lambda y - z\| \geq d = \|y_0 - z\| = \|x\|.$$

So

$$\|x + \lambda y\|^2 - \|x\|^2 \geq 0.$$

Now,

$$\|x + \lambda y\|^2 - \|x\|^2 = \overline{\lambda}(x, y) + \lambda(y, x) + |\lambda|^2 \|x\|^2.$$

Let $\lambda = \mu(x, y)$, where $\mu \in \mathbb{R}$ is to be suitably chosen. Hence,

$$2\mu|(x, y)|^2 + \mu^2 |(x, y)|^2 \|y\|^2 \geq 0.$$

If $\mu$ is negative and $|\mu|$ is small, then the linear term in $\mu$ dominates on the left-hand side of the above inequality. This can happen only if $(x, y) = 0$, i.e., $x \perp y$, for every $y \in N$, i.e., $x \perp N$, completing the proof. □

**Theorem 74.** *If $M$ is a closed subspace of $H$, then*

$$M = (M^\perp)^\perp.$$

*Proof.* Suppose $M \neq (M^\perp)^\perp$. Then, by Theorem 73, there exists $x \in (M^\perp)^\perp$, $x \neq 0$ such that $x \perp M$, i.e., $x \in M^\perp$. This implies that

$$x \in M^\perp \cap (M^\perp)^\perp = \{0\},$$

and so $x = 0$, which is a contradiction, proving the result. □

Suppose $\{M_a\}_{a\in\Lambda}$ is a family of subspaces of $H$. Then $\{M_a^\perp\}_{a\in\Lambda}$ is a family of closed subspaces of $H$ and so $\bigcap_{a\in\Lambda} M_a^\perp$ is a closed subspace of $H$. On the other hand, we have the following result:

**Lemma 5.** *Let $M$ be a non-empty subset of $H$. Then the closed subspace generated by $M$ in $H$ is the closure of the set of all finite linear combinations of elements of $M$.*

*Proof.* The proof of the lemma follows immediately because the set of all finite linear combinations of elements of $M$ is precisely the linear subspace spanned by $M$. The details are left to the reader. □

If $M$ is the closed subspace generated by $\bigcup_{a\in\Lambda} M_a$, then

$$M^\perp = \bigcap_{a\in\Lambda} M_a^\perp.$$

In a similar fashion, we can see that $\left(\bigcap_{a\in\Lambda} M_a\right)^\perp$ is the closed subspace generated by $\bigcup_{a\in\Lambda} M_a^\perp$.

**Theorem 75.** *If $M, N$ are closed subspaces of $H$ such that $M \perp N$, then their sum $M + N$ is also a closed subspace of $H$.*

*Proof.* Let $\{x_n\}$ be a sequence in $M + N$ such that

$$x_n \to x, \; n \to \infty \text{ in } H.$$

Since $x_n \in M + N$, $n = 1, 2, \ldots$, we can write

$$x_n = x_n' + x_n'', \; x_n' \in M, \; x_n'' \in N, \; n = 1, 2, \ldots.$$

Since $\{x_n\}$ converges, it is a Cauchy sequence. Since $M \perp N$,

$$\|x_n - x_m\|^2 = \|x_n' - x_m'\|^2 + \|x_n'' - x_m''\|^2.$$

It now follows that $\{x'_n\}$ is a Cauchy sequence in $M$ and $\{x''_n\}$ is a Cauchy sequence in $N$. So

$$x'_n \to x',\ x''_n \to x,''\ \ n \to \infty \text{ in } H.$$

Using the fact that $M$ and $N$ are closed, $x' \in M$, $x'' \in N$. Now, given $\epsilon > 0$, we can find a positive integer $n_0$ such that, for all $n \geq n_0$, we have

$$\|x_n - x\| < \frac{\epsilon}{3},$$
$$\|x'_n - x'\| < \frac{\epsilon}{3}$$

and

$$\|x''_n - x''\| < \frac{\epsilon}{3}.$$

Hence, for $n \geq n_0$,

$$\begin{aligned}\|x - (x' + x'')\| &= \|(x - x_n) + (x'_n - x') + (x''_n - x'') \\ &\qquad + \{x_n - (x'_n + x''_n)\}\| \\ &= \|(x - x_n) + (x'_n - x') + (x''_n - x'')\|, \\ &\qquad \text{since } x_n = x'_n + x''_n \\ &< \frac{\epsilon}{3} + \frac{\epsilon}{3} + \frac{\epsilon}{3} \\ &= \epsilon.\end{aligned}$$

Since $\epsilon > 0$ is arbitrary,

$$x = x' + x,'' \text{ where } x' \in M, x'' \in N$$

and so $x \in M + N$. Thus $M + N$ is closed, proving the result. □

We now prove an important result, known as the projection theorem.

**Theorem 76** (Projection theorem). *If $M$ is a closed subspace of a Hilbert space $H$, then*

$$H = M \oplus M^{\perp}.$$

*Proof.* Since $M \perp M^\perp$, by Theorem 75, $M + M^\perp$ is a closed subspace of $H$. In view of Theorem 71 (iii), since $M \subset M + M^\perp$,

$$M^\perp \supset (M + M^\perp)^\perp$$

and similarly,

$$(M^\perp)^\perp \supset (M + M^\perp)^\perp$$

and so

$$M^\perp \cap (M^\perp)^\perp \supset (M + M^\perp)^\perp.$$

But

$$M^\perp \cap (M^\perp)^\perp = \{0\}$$

and so

$$(M + M^\perp)^\perp = \{0\}.$$

Now,

$$M + M^\perp = ((M + M^\perp)^\perp)^\perp,$$

using Theorem 74, since $M + M^\perp$ is a closed subspace of $H$. Hence,

$$M + M^\perp = \{0\}^\perp = H.$$

We now observe that $M \cap M^\perp = \{0\}$, to conclude that

$$H = M \oplus M^\perp,$$

completing the proof.
Alternate Proof. Since $M, M^\perp$ are orthogonal closed subspaces of $H$, $M + M^\perp$ is also a closed linear subspace of $H$. We claim that $M + M^\perp = H$. Suppose not. Then, using Theorem 73, there exists a vector $z_0 \neq 0$ such that

$$z_0 \perp (M + M^\perp).$$

So $z_0 \in M^\perp \cap (M^\perp)^\perp$ and $z_0 \neq 0$, which is not possible. Thus,

$$H = M + M^\perp.$$

It follows, as before, that

$$H = M \oplus M^{\perp}.$$

The proof is complete. □

---

## 7.3 Infinite sums

**Theorem 77.** *If $\{e_i\}_{i \in \Lambda}$ is an orthonormal set in a Hilbert space $H$ and $x$ is any vector in $H$, then the set*

$$S = \{e_i / (x, e_i) \neq 0\}$$

*is either empty or countable.*

*Proof.* For $n = 1, 2, \ldots$, let

$$S = \left\{ e_i / |(x, e_i)|^2 > \frac{\|x\|^2}{n} \right\}.$$

By Theorem 70, $S_n$ contains at most $(n-1)$ vectors. Noting that

$$S = \bigcup_{n=1}^{\infty} S_n,$$

the conclusion follows. □

Let $\{e_i\}_{i \in \Lambda}$ be an orthonormal set in a Hilbert space $H$. We define the sum $\sum_{i \in \Lambda} |(x, e_i)|^2$ as follows:
Let

$$S = \{e_i / (x, e_i) \neq 0\}.$$

If $S = \phi$, we define $\sum_{i \in \Lambda} |(x, e_i)|^2$ to be 0. If $S \neq \phi$, $S$ is finite or countably infinite. If $S$ is finite, it can be written in the form

$$S = \{e_1, e_2, \ldots, e_n\},$$

for a positive integer $n$. In this case, we define $\sum_{i\in\Lambda}|(x,e_i)|^2$ to be $\sum_{i=1}^{n}|(x,e_i)|^2$, which is independent of the order in which the elements of $S$ are arranged. It remains to consider the case when $S$ is countably infinite. So we can arrange the vectors in $S$ as

$$\{e_1, e_2, \ldots, e_n, \ldots\}$$

in some order. We know that if $\sum_{i=1}^{\infty}|(x,e_i)|^2$ converges, then every series obtained from the above series by rearranging its terms also converges to the same sum. So, in this case, we define $\sum_{i\in\Lambda}|(x,e_i)|^2$ to be $\sum_{i=1}^{\infty}|(x,e_i)|^2$.

Having understood the meaning of the sum $\sum_{i\in\Lambda}|(x,e_i)|^2$, we now have the following result.

**Theorem 78** (Generalized Bessel's inequality). *If $\{e_i\}_{i\in\Lambda}$ is an orthonormal set in a Hilbert space $H$, then*

$$\sum_{i\in\Lambda}|(x,e_i)|^2 \le \|x\|^2, \text{ for every } x \in H.$$

*Proof.* Let

$$S = \{e_i/(x,e_i) \neq 0\}.$$

By Theorem 77, $S = \phi$ or $S$ is finite or $S$ is countably infinite.

Case 1. If $S = \phi$,

$$\sum_{i\in\Lambda}|(x,e_i)|^2 = 0$$

and the result holds trivially.

Case 2. If $S$ is finite, then

$$S = \{e_1, e_2, \ldots, e_n\}, \text{ for some positive integer } n.$$

In this case,

$$\sum_{i\in\Lambda}|(x,e_i)|^2 = \sum_{i=1}^{n}|(x,e_i)|^2$$

and the result follows from Theorem 70.

Case 3. $S$ is countably infinite. In this case,

$$\sum_{i\in\Lambda} |(x, e_i)|^2 = \sum_{i=1}^{\infty} |(x, e_i)|^2.$$

Using Theorem 70,

$$\sum_{i=1}^{n} |(x, e_i)|^2 \leq \|x\|^2, \text{ for every } n = 1, 2, \ldots,$$

from which it follows that

$$\sum_{i=1}^{\infty} |(x, e_i)|^2 \leq \|x\|^2,$$

completing the proof of the theorem. □

We note the following about a Hilbert space $H$. If $H$ contains only the zero vector, it has no orthonormal sets. If $H \neq \{0\}$, then there exists $x \in H$, $x \neq 0$. Consider $e = \frac{x}{\|x\|}$. Then $\|e\| = 1$ and the singleton set $\{e\}$ is an orthonormal set in $H$. More generally, if $(x_i)_{i\in\Lambda}$ is a non-empty set of mutually orthogonal, non-zero vectors in $H$, then $(e_i)_{i\in\Lambda}$, where

$$e_i = \frac{x_i}{\|x_i\|}, \text{ for } i \in \Lambda,$$

is an orthonormal set in $H$.

We now have the following result.

**Theorem 79.** *Every Hilbert space $H \neq \{0\}$ contains a complete orthonormal set, in the sense that $H$ contains a maximal orthonormal set.*

*Proof.* Since $H \neq \{0\}$, the class of all its orthonormal sets is non-empty. This class is a partially ordered set under "set inclusion." Since the union of the elements of any chain of orthonormal sets is an upper bound for the chain in the partially ordered set of all orthonormal sets, the theorem follows immediately from Zorn's lemma. □

Complete orthonormal sets are truly interesting, due to the following result.

**Theorem 80.** *Let $H$ be a Hilbert space and let $(e_i)_{i\in\Lambda}$ be an orthonormal set in $H$. Then the following statements are equivalent:*

*(i)* $(e_i)_{i\in\Lambda}$ *is complete;*

*(ii)* $x \perp e_i$ *for every* $i \in \Lambda$ *implies that* $x = 0$*;*

*(iii) if $x$ is any vector in $H$, then* $x = \sum_{i\in\Lambda}(x, e_i)e_i$*;*

*and*

*(iv) if $x$ is any vector in $H$, then* $\|x\|^2 = \sum_{i\in\Lambda}|(x, e_i)|^2$.

*Proof.* (i) implies (ii). Let (i) hold. Suppose (ii) does not hold. So there exists $x \neq 0$ in $H$ such that $x \perp e_i$, for every $i \in \Lambda$. Then $(e, e_i)_{i\in\Lambda}$ is an orthonormal set, which properly contains $(e_i)_{i\in\Lambda}$, which contradicts (i). Thus (i) implies (ii);

(ii) implies (iii). Note that $x - \sum_{i\in\Lambda}(x, e_i)e_i \perp e_i$, for every $i \in \Lambda$. In view of (ii),

$$x - \sum_{i\in\Lambda}(x, e_i)e_i = 0,$$

$$\text{i.e.,} \quad x = \sum_{i\in\Lambda}(x, e_i)e_i,$$

so that (iii) holds, i.e., (ii) implies (iii);

(iii) implies (iv). Let (iii) hold. Then

$$\begin{aligned}\|x\|^2 = (x, x) &= \left(\sum_{i\in\Lambda}(x, e_i)e_i, \sum_{i\in\Lambda}(x, e_i)e_i\right) \\ &= \sum_{i\in\Lambda}(x, e_i)\overline{(x, e_i)} \\ &= \sum_{i\in\Lambda}|(x, e_i)|^2,\end{aligned}$$

so that (iv) holds, proving that (iii) implies (iv);

(iv) implies (i). Let (iv) hold. Suppose $(e_i)_{i\in\Lambda}$ is not complete, it is a proper subset of an orthonormal set, (say) $(e, e_i)_{i\in\Lambda}$. $e \perp e_i$, for all $i \in \Lambda$ so that

$(e, e_i) = 0$, for all $i \in \Lambda$. Because of (iv), we now have

$$\|e\|^2 = \sum_{i \in \Lambda} |(e, e_i)|^2 = 0,$$

which implies $e = 0$, contradicting the fact that $\|e\| = 1$. This contradiction proves that (iv) implies (i) and the proof of the theorem is now complete. □

**Definition 77.** *Let $(e_i)_{i \in \Lambda}$ be a complete, orthonormal set in a Hilbert space $H$ and $x$ be any vector in $H$. The numbers $(x, e_i)$, $i \in \Lambda$ are called the "Fourier coefficients" of $x$, the expression*

$$x = \sum_{i \in \Lambda} (x, e_i) e_i$$

*is called the "Fourier expansion" of $x$. The equation*

$$\|x\|^2 = \sum_{i \in \Lambda} |(x, e_i)|^2$$

*is called "Parseval's equation," all with respect to the complete orthonormal set $(e_i)_{i \in \Lambda}$ under consideration.*

---

## 7.4 Riesz representation theorem

We now make some important observations:
Let $y$ be a fixed vector in a Hilbert space $H$. Define the mapping $f_y$ on $H$ by:

$$f_y(x) = (x, y), \text{ for all } x \in H.$$

Now, for $x_1, x_2 \in H$ and scalars $\alpha, \beta$,

$$\begin{aligned} f_y(\alpha x_1 + \beta x_2) &= (\alpha x_1 + \beta x_2, y) \\ &= \alpha (x_1, y) + \beta (x_2, y) \\ &= \alpha f_y(x_1) + \beta f_y(x_2), \end{aligned}$$

proving that $f_y$ is linear. Also,

$$\begin{aligned} |f_y(x)| &= |(x, y)| \\ &\leq \|x\| \|y\|, \end{aligned}$$

using Schwarz inequality. Thus, $f_y$ is bounded and

$$\|f_y\| \leq \|y\|.$$

We claim that

$$\|f_y\| = \|y\|.$$

This is clear if $y = 0$. If $y \neq 0$,

$$\begin{aligned} \|f_y\| &= \sup_{\|x\|=1} |f_y(x)| \\ &\geq \left| f_y\left(\frac{y}{\|y\|}\right) \right| \\ &= \left| \left(\frac{y}{\|y\|}, y\right) \right| \\ &= \frac{1}{\|y\|} |(y, y)| \\ &= \frac{1}{\|y\|} \|y\|^2 \\ &= \|y\|, \end{aligned}$$

proving the claim. Thus, the mapping $y \to f_y$ is a norm-preserving mapping of $H$ into $\tilde{H}$. In the context of this result, the interesting fact is that every bounded linear functional on $H$ arises precisely in this way.

**Theorem 81** (Riesz representation theorem). *Let $H$ be a Hilbert space and let $f \in \tilde{H}$. Then there exists a unique vector $y \in H$ such that*

$$f(x) = (x, y), \text{ for all } x \in H. \tag{7.1}$$

*Proof.* First, we prove that if such a vector $y \in H$ exists, it is unique. Suppose $y' \in H$ also satisfies the requirements of the theorem, i.e.,

$$f(x) = (x, y'), \text{ for all } x \in H.$$

Then

$$(x, y') = (x, y), \text{ for all } x \in H,$$
$$i.e., \ (x, y' - y) = 0, \text{ for all } x \in H.$$

Choosing $x = y' - y$, we get

$$(y' - y, y' - y) = 0,$$
$$i.e., \ \|y' - y\|^2 = 0,$$
$$i.e., \ \|y' - y\| = 0,$$
$$i.e., \ y' = y,$$

proving the uniqueness of $y$. We now claim the existence of such a vector $y$ in $H$. If $f = 0$, choose $y = 0$ and we are done in this case. So, let us now suppose that $f \neq 0$. Hence the null space $M$ of $f$ is a proper, closed, linear subspace of $H$. In view of Theorem 73, there exists $y_0 \neq 0$, which is orthogonal to $M$. We claim that for a suitably chosen scalar $\alpha$, $y = \alpha y_0$ meets our requirements. We first observe that (7.1) holds, for every $x \in M$ trivially, because if $x \in M$, then $f(x) = 0$. Since $y_0 \perp x$, $(x, y) = (x, \alpha y_0) = \overline{\alpha}(x, y_0) = 0$. We now choose $\alpha$ such that (7.1) holds for $x = y_0$, i.e.,

$$f(y_0) = (y_0, y),$$
$$i.e., \ f(y_0) = (y_0, \alpha y_0)$$
$$= \overline{\alpha}(y_0, y_0)$$
$$= \overline{\alpha}\|y_0\|^2,$$
$$i.e., \ \overline{\alpha} = \frac{f(y_0)}{\|y_0\|^2},$$
$$i.e., \ \alpha = \frac{\overline{f(y_0)}}{\|y_0\|^2},$$

so that (7.1) holds for every $x \in M$ and for $x = y_0$ also. Note that every $x \in H$ can be written as

$$x = m + \beta y_0, \ m \in M \text{ and for a suitable scalar } \beta.$$

To do this, choose $\beta$ such that

$$f(x - \beta y_0) = 0,$$

$$i.e., \quad f(x) - \beta f(y_0) = 0,$$

$$i.e., \quad \beta = \frac{f(x)}{f(y_0)}.$$

Now,

$$\begin{aligned} f(x) &= f(m + \beta y_0) \\ &= f(m) + \beta f(y_0) \\ &= (m, y) + \beta(y_0, y) \\ &= (m + \beta y_0, y) \\ &= (x, y), \end{aligned}$$

completing the proof of the theorem. □

**Remark 5.** *The above result shows that the mapping*

$$y \to f_y, \text{ where } f_y(x) = (x, y) \tag{7.2}$$

*is a norm-preserving mapping of $H$ onto $\tilde{H}$. The above mapping is not linear, since*

$$f_{\alpha y} = \overline{\alpha} f_y.$$

*However, the mapping (7.2) is an isometry, because*

$$\|f_x - f_y\| = \|f_{x-y}\| = \|x - y\|.$$

## Exercises

1. If $\{x_1, x_2, \ldots, x_n\}$ is a finite orthogonal family, prove that

$$\left\| \sum_{i=1}^{n} x_i \right\|^2 = \sum_{i=1}^{n} \|x_i\|^2.$$

2. Deduce Schwarz inequality from Bessel's inequality.

3. Prove that every finite-dimensional subspace of a Hilbert space has an orthogonal complement.

4. With regard to the mapping in (7.2), verify the relations

$$f_{y_1+y_2} = f_{y_1} + f_{y_2}; f_{\alpha y} = \overline{\alpha} f_y.$$

5. Let $H$ be a Hilbert space. Prove that $\tilde{H}$ is also a Hilbert space with respect to the inner product defined by

$$(f_x, f_y) = (y, x).$$

Prove that $\tilde{\tilde{H}}$ is also a Hilbert space with regard to the inner product defined by

$$(F_f, F_g) = (g, f).$$

6. Prove that a closed, convex subset of a Hilbert space $H$ contains a unique vector of smallest norm.

7. If $M$ is a linear subspace of a Hilbert space $H$, prove that $M$ is closed if and only if $M = (M^{\perp})^{\perp}$.

8. Let $S$ be a non-empty subset of a Hilbert space $H$. Prove that the set of all linear combinations of vectors in $S$ is dense in $H$ if and only if $S^{\perp} = \{0\}$.

9. Let $\{e_1, e_2, \ldots, e_n\}$ be a finite orthonormal set in a Hilbert space $H$. If $x$ is any vector in $H$, prove that

$$x - \sum_{i=1}^{n} (x, e_i)e_i \perp e_j, \text{ for every } j = 1, 2, \ldots, n.$$

10. Extend the result in Exercise 9 to any arbitrary orthonormal set $\{e_i\}_{i \in \Lambda}$ in a Hilbert space $H$.

11. Prove that every orthonormal set in a Hilbert space $H$ is contained in some complete, orthonormal set in $H$.

12. Prove that a Hilbert space $H$ is separable if and only if every orthonormal set in $H$ is countable.

13. Prove that any two complete, orthonormal sets in a Hilbert space $H$ have the same cardinal number. (This number is called the orthogonal dimension of $H$. In case, if $H$ has no complete, orthonormal sets, its orthogonal dimension is defined to be 0).

14. Prove that a separable, infinite-dimensional Hilbert space $H$ is isomorphic to $\ell_2$.

15. Let $M_1, M_2$ be subspaces of an inner product space $X$. Prove that $M_1 \perp M_2$ if and only if

$$\|m_1 + m_2\|^2 = \|m_1\|^2 + \|m_2\|^2,$$

whenever, $m_1 \in M_1$ and $m_2 \in M_2$.

16. Let $G$ be a closed subspace of a Hilbert space $H$. Use $H = G \oplus G^{\perp}$, to prove that $(G^{\perp})^{\perp} = G$.

# Chapter 8

# Silverman–Toeplitz Theorem and Schur's Theorem

## 8.1 Silverman–Toeplitz theorem

In this section, we study matrix transformations between sequence spaces and we prove a landmark result in summability theory, viz., the Silverman–Toeplitz theorem.

The motivation for summability theory is divergent series. First, consider the sequence

$$\{x_n\} = \{1, 0, 1, 0, \dots\},$$

which is known to diverge. However, let

$$t_n = \frac{x_0 + x_1 + \cdots + x_n}{n+1}, \; n = 0, 1, 2, \dots,$$

$$i.e., \; t_n = \begin{cases} \frac{k+1}{2k+1}, & \text{if } n = 2k; \\ \frac{k+1}{2k+2}, & \text{if } n = 2k+1, \end{cases}$$

so that

$$t_n \to \frac{1}{2}, \; n \to \infty.$$

In this case, we say that the sequence $\{x_n\}$ converges to $\frac{1}{2}$ (in fact, summable to $\frac{1}{2}$) in the sense of Cesàro or $\{x_n\}$ is $(C, 1)$ summable to $\frac{1}{2}$. Similarly, consider the infinite series

$$\sum_{n=0}^{\infty} a_n = 1 - 1 + 1 - 1 + \cdots.$$

The associated sequence $\{s_n\}$ of partial sums is $\{1, 0, 1, 0, \dots\}$, which is $(C, 1)$ summable to $\frac{1}{2}$. In this case, we say that the series $1 - 1 + 1 - 1 + \cdots$ is $(C, 1)$ summable to $\frac{1}{2}$. We now introduce the following definitions and concepts.

Hereafter, entries of sequences, series, and infinite matrices are real or complex numbers.

**Definition 78.** *Given an infinite matrix $A = (a_{nk})$, $n, k = 0, 1, 2, \dots$ and a sequence $x = \{x_k\}$, $k = 0, 1, 2, \dots$, by the A-transform of $x = \{x_k\}$, we mean the sequence $A(x) = \{(Ax)_n\}$,*

$$(Ax)_n = \sum_{k=0}^{\infty} a_{nk} x_k, \ n = 0, 1, 2, \dots,$$

*where we suppose that the series on the right converge. If $\lim_{n \to \infty} (Ax)_n = s$, we say that the sequence $x = \{x_k\}$ is A-summable or summable A to s.*

Given the infinite series $\sum_{k=0}^{\infty} x_k$, we define

$$s_n = \sum_{k=0}^{n} x_k, \ n = 0, 1, 2, \dots.$$

If the sequence $\{s_n\}$ is $A$-summable to $s$, we say that $\sum_{k=0}^{\infty} x_k$ is $A$-summable to $s$.

**Definition 79.** *Let $X, Y$ be sequence spaces. The infinite matrix $A = (a_{nk})$, $n, k = 0, 1, 2, \dots$ is said to transform $X$ to $Y$, written as $A \in (X, Y)$, if whenever the sequence $x = \{x_k\} \in X$, $(Ax)_n$ is defined, for every $n = 0, 1, 2, \dots$ and the sequence $\{(Ax)_n\} \in Y$.*

**Definition 80.** *Let $c$, as usual, be the Banach space of all convergent sequences under the norm*

$$\|x\| = \sup_{k \geq 0} |x_k|, \ x = \{x_k\} \in c.$$

*If $A = (a_{nk}) \in (c, c)$, $A$ is said to be "convergence-preserving" or "conservative." If $A$ is conservative and in addition,*

$$\lim_{n \to \infty} (Ax)_n = \lim_{k \to \infty} x_k, \ x = \{x_k\} \in c,$$

*A is called a "regular" matrix or a "Toeplitz" matrix. If A is regular, we write*

$$A \in (c, c; P),$$

*where the letter P denotes "preservation of limit."*

We now have the following landmark result, due to Silverman–Toeplitz, which characterizes a regular matrix in terms of the entries of the matrix.

**Theorem 82** (Silverman–Toeplitz). *$A = (a_{nk}) \in (c, c; P)$, i.e., $A$ is regular if and only if*

$$(i) \ \sup_{n\geq 0} \sum_{k=0}^{\infty} |a_{nk}| < \infty; \tag{8.1}$$

$$(ii) \ \lim_{n\to\infty} a_{nk} = 0, \ k = 0, 1, 2, \ldots; \tag{8.2}$$

*and*

$$(iii) \ \lim_{n\to\infty} \sum_{k=0}^{\infty} a_{nk} = 1. \tag{8.3}$$

*Proof.* Sufficiency part. Let (8.1), (8.2), and (8.3) hold. Let $x = \{x_k\} \in c$ with $\lim_{k\to\infty} x_k = s$. Since $\{x_k\}$ converges, it is bounded and so

$$|x_k| \leq M, \ M > 0, \ k = 0, 1, 2, \ldots.$$

Now,

$$\begin{aligned} \sum_{k=0}^{\infty} |a_{nk} x_k| &\leq M \sum_{k=0}^{\infty} |a_{nk}| \\ &\leq M \sup_{n\geq 0} \sum_{k=0}^{\infty} |a_{nk}| \\ &< \infty, \text{ in view of (8.1)} \end{aligned}$$

and so $\sum_{k=0}^{\infty} a_{nk} x_k$ converges absolutely, $n = 0, 1, 2, \ldots$. Thus,

$$(Ax)_n = \sum_{k=0}^{\infty} a_{nk} x_k \text{ is defined}, n = 0, 1, 2, \ldots.$$

Now we have

$$(Ax)_n = \sum_{k=0}^{\infty} a_{nk}(x_k - s) + s\sum_{k=0}^{\infty} a_{nk},\ n = 0, 1, 2, \ldots. \tag{8.4}$$

Since $\lim_{k\to\infty} x_k = s$, given $\epsilon > 0$, there exists a positive integer $N$ such that

$$|x_k - s| < \frac{\epsilon}{2L},\ \text{for all } k > N, \tag{8.5}$$

where $L > 0$ is such that

$$|x_n - s| \leq L,\ \sum_{k=0}^{\infty} |a_{nk}| \leq L,\ n = 0, 1, 2, \ldots. \tag{8.6}$$

Consequently,

$$\sum_{k=0}^{\infty} a_{nk}(x_k - s) = \sum_{k=0}^{N} a_{nk}(x_k - s) + \sum_{k=N+1}^{\infty} a_{nk}(x_k - s),$$

$$\left|\sum_{k=0}^{\infty} a_{nk}(x_k - s)\right| \leq \sum_{k=0}^{N} |a_{nk}||x_k - s| + \sum_{k=N+1}^{\infty} |a_{nk}||x_k - s|.$$

$$\begin{aligned}\sum_{k=N+1}^{\infty} |a_{nk}||x_k - s| &< \frac{\epsilon}{2L}\sum_{k=0}^{\infty} |a_{nk}|,\ \text{using (8.5)}\\ &< \frac{\epsilon}{2L}L,\ \text{using (8.6)}\\ &= \frac{\epsilon}{2};\end{aligned}$$

Using (8.2), we can find a positive integer $N'$ such that

$$|a_{nk}| < \frac{\epsilon}{2L(N+1)},\ k = 0, 1, 2, \ldots, N,\ \text{for all } n > N'$$

and so

$$\begin{aligned}\sum_{k=0}^{N} |a_{nk}||x_k - s| &< (N+1)L\frac{\epsilon}{2L(N+1)}\\ &= \frac{\epsilon}{2},\ \text{for all } n > N'.\end{aligned}$$

Thus,

$$\begin{aligned}\left|\sum_{k=0}^{\infty} a_{nk}(x_k - s)\right| &< \frac{\epsilon}{2} + \frac{\epsilon}{2}\\ &= \epsilon,\ \text{for all } n > N',\end{aligned}$$

so that

$$\lim_{n\to\infty}\sum_{k=0}^{\infty} a_{nk}(x_k - s) = 0. \tag{8.7}$$

Now, taking the limit as $n \to \infty$ in (8.4), we get

$$\lim_{n\to\infty}(Ax)_n = s, \text{ using (8.3) and (8.7),}$$

and so $A$ is regular.

Necessity part. Let $A$ be regular. For every fixed $k = 0, 1, 2, \ldots$, consider the sequence $x = \{x_n\}$, where

$$x_n = \begin{cases} 1, & \text{if } n = k; \\ 0, & \text{otherwise.} \end{cases}$$

For this sequence $x$,

$$(Ax)_n = a_{nk}.$$

Since $\lim_{n\to\infty} x_n = 0$ and $A$ is regular, it follows that

$$\lim_{n\to\infty} a_{nk} = 0, \ k = 0, 1, 2, \ldots,$$

so that (8.2) holds. Again, consider the sequence $x = \{x_n\}$, where $x_n = 1$, $n = 0, 1, 2, \ldots$. Note that $\lim_{n\to\infty} x_n = 1$. For this sequence $x$,

$$(Ax)_n = \sum_{k=0}^{\infty} a_{nk}.$$

Since $\lim_{n\to\infty} x_n = 1$ and $A$ is regular, we get,

$$\lim_{n\to\infty}\sum_{k=0}^{\infty} a_{nk} = 1,$$

so that (8.3) holds. It remains to prove that (8.1) holds. First, we prove that

$$\sum_{k=0}^{\infty} |a_{nk}| \text{ converges, for every } n = 0, 1, 2, \ldots.$$

Suppose not. Then there exists a positive integer $N$ such that

$$\sum_{k=0}^{\infty} |a_{Nk}| \text{ diverges.}$$

In fact, $\sum_{k=0}^{\infty} |a_{Nk}|$ diverges to $\infty$. So we can find a strictly increasing sequence $\{k(j)\}$ of positive integers such that

$$\sum_{k=k(j-1)}^{k(j)-1} |a_{Nk}| > 1, \; j = 1, 2, \ldots. \tag{8.8}$$

Define the sequence $x = \{x_k\}$, where

$$x_k = \begin{cases} \frac{|a_{Nk}|}{ja_{Nk}}, & \text{if } a_{Nk} \neq 0 \text{ and } k(j-1) \leq k < k(j), j = 1, 2, \ldots; \\ 0, & \text{if } k = 0 \text{ or } a_{Nk} = 0. \end{cases}$$

Note that $\lim_{k\to\infty} x_k = 0$ and $\sum_{k=0}^{\infty} a_{nk}x_k$ converges, $n = 0, 1, 2, \ldots$. In particular, $\sum_{k=0}^{\infty} a_{Nk}x_k$ converges. However,

$$\begin{aligned} \sum_{k=0}^{\infty} a_{Nk}x_k &= \sum_{j=1}^{\infty} \sum_{k=k(j-1)}^{k(j)-1} \frac{|a_{Nk}|}{j} \\ &= \sum_{j=1}^{\infty} \frac{1}{j} \sum_{k=k(j-1)}^{k(j)-1} |a_{Nk}| \\ &> \sum_{j=1}^{\infty} \frac{1}{j}. \end{aligned}$$

Since $\sum_{j=1}^{\infty} \frac{1}{j}$ diverges, $\sum_{k=0}^{\infty} a_{Nk}x_k$ diverges, which is a contradiction. Thus,

$$\sum_{k=0}^{\infty} |a_{nk}| \text{ converges, for every } n = 0, 1, 2, \ldots.$$

Now, suppose (8.1) does not hold, i.e.,

$$\sup_{n\geq 0} \sum_{k=0}^{\infty} |a_{nk}| = \infty.$$

We arrive at a contradiction.

We construct two strictly increasing sequences $\{m(j)\}$ and $\{n(j)\}$ of positive integers as follows:

Let $m(0) = 0$. Since $\sum_{k=0}^{\infty} |a_{m(0),k}|$ converges, choose a positive integer $n(0)$ such that

$$\sum_{k=n(0)+1}^{\infty} |a_{m(0),k}| < 1.$$

Having chosen the positive integers $m(0), m(1), \ldots, m(j-1)$ and $n(0), n(1), \ldots, n(j-1)$, choose the positive integers $m(j) > m(j-1)$ and $n(j) > n(j-1)$ such that

$$\sum_{k=0}^{\infty} |a_{m(j),k}| > j^2 + 2j + 2; \tag{8.9}$$

$$\sum_{k=0}^{n(j-1)} |a_{m(j),k}| < 1; \tag{8.10}$$

and

$$\sum_{k=n(j)+1}^{\infty} |a_{m(j),k}| < 1. \tag{8.11}$$

Now, define the sequence $x = \{x_k\}$, where

$$x_k = \begin{cases} \frac{|a_{m(j),k}|}{j a_{m(j),k}}, & \text{if } a_{m(j),k} \neq 0, n(j-1) < k \leq n(j), j = 1, 2, \ldots; \\ 0, & \text{otherwise.} \end{cases}$$

Note that $\lim_{k\to\infty} x_k = 0$. Since $A$ is regular, $\lim_{n\to\infty} (Ax)_n = 0$. However,

$$\begin{aligned}
|(Ax)_{m(j)}| &= \left|\sum_{k=0}^{\infty} a_{m(j),k} x_k\right| \\
&= \left|\sum_{k=0}^{n(j-1)} a_{m(j),k} x_k + \sum_{k=n(j-1)+1}^{n(j)} a_{m(j),k} x_k \right. \\
&\qquad \left. + \sum_{k=n(j)+1}^{\infty} a_{m(j),k} x_k\right| \\
&\geq \left|\sum_{k=n(j-1)+1}^{n(j)} a_{m(j),k} x_k\right| - \sum_{k=0}^{n(j-1)} |a_{m(j),k} x_k| \\
&\qquad - \sum_{k=n(j)+1}^{\infty} |a_{m(j),k} x_k| \\
&> \frac{1}{j} \sum_{k=n(j-1)+1}^{n(j)} |a_{m(j),k}| - 1 - 1, \text{ using (8.10) and (8.11)} \\
&= \frac{1}{j}\left[\sum_{k=0}^{\infty} |a_{m(j),k}| - \sum_{k=0}^{n(j-1)} |a_{m(j),k}| \right. \\
&\qquad \left. - \sum_{k=n(j)+1}^{\infty} |a_{m(j),k}|\right] - 2 \\
&> \frac{1}{j}[j^2 + 2j + 2 - 1 - 1] - 2, \text{ using (8.9), (8.10), (8.11)} \\
&= j + 2 - 2 \\
&= j, \ j = 1, 2, \ldots .
\end{aligned}$$

Thus, $\{(Ax)_{m(j)}\}$ diverges, contradicting the fact that $\{(Ax)_n\}$ converges. Consequently, (8.1) holds, completing the proof of the theorem. □

One can prove the following result on the same lines as above, the details of which are left to the reader.

**Theorem 83** (Kojima–Schur). *$A = (a_{nk}) \in (c, c)$, i.e., $A$ is conservative if and only if (8.1) holds and*

$$\lim_{n\to\infty} a_{nk} = \delta_k, \ k = 0, 1, 2, \ldots; \tag{8.12}$$

*and*

$$\lim_{n\to\infty} \sum_{k=0}^{\infty} a_{nk} = \delta. \tag{8.13}$$

*In such a case,*

$$\lim_{n\to\infty} (Ax)_n = s\delta + \sum_{k=0}^{\infty} (x_k - s)\delta_k,$$

*where $\lim_{k\to\infty} x_k = s$.*

**Remark 6.** *Whatever the case, $A$ is conservative or regular, $A$ is a bounded linear mapping of $c$ into $c$ and*

$$\|A\| = \sup_{n\geq 0} \sum_{k=0}^{\infty} |a_{nk}|.$$

---

## 8.2 Examples of regular methods

**Example 27.** *The arithmetic means. The $(C, 1)$ matrix defined by*

$$a_{nk} = \begin{cases} \frac{1}{n+1}, & \text{if } 0 \leq k \leq n; \\ 0, & \text{if } k > n, \end{cases}$$

*is a regular matrix. The $(C, 1)$ matrix is called the Cesàro matrix of order 1.*

**Example 28.** *Cesàro means of order $r$. For each $r > -1$, the $(C, r)$ matrix, called the Cesàro matrix of order $r$, is defined by*

$$a_{nk} = \begin{cases} \frac{A^{r-1}_{n-k}}{A^r_n}, & \text{if } 0 \leq k \leq n; \\ 0, & \text{if } k > n, \end{cases}$$

*where*

$$A_0^r = 1;$$

$$A_n^r = \frac{(r+1)(r+2)\dots(r+n)}{n!}, \text{ for all } n = 1, 2, \dots.$$

*When* $r \geq 0$,

$$A_n^{r-1} \geq 0$$

*and using the fact that*

$$(1-z)^{-(r+1)} = \sum_{n=0}^{\infty} A_n^r z^n, \ |z| < 1,$$

*one can prove that*

$$\sum_{k=0}^{n} A_{n-k}^s A_k^r = A_n^{s+r+1},$$

*from which it follows that*

$$\sum_{k=0}^{\infty} |a_{nk}| = \sum_{k=0}^{\infty} a_{nk} = 1, \text{ for every } n = 0, 1, 2, \dots,$$

*from which we get*

$$\sup_{n \geq 0} \sum_{k=0}^{\infty} |a_{nk}| < \infty$$

*and*

$$\lim_{n \to \infty} \sum_{k=0}^{\infty} a_{nk} = 1.$$

*Using elementary analysis, one can see that*

$$\frac{A_n^r}{n^r} \to \frac{1}{\gamma(r+1)}, \ n \to \infty,$$

*where* $\gamma(x)$ *denotes the gamma function. It now follows that*

$$\lim_{n \to \infty} a_{nk} = 0, \ k = 0, 1, 2, \dots.$$

*Thus, we have shown that the* $(C, r)$ *matrix is a regular matrix, when* $r \geq 0$. *In this context, note that the case* $r = 0$ *gives the unit matrix and the case* $r = 1$ *gives the arithmetic mean. We also note that the* $(C, r)$ *matrix need not be regular if* $-1 < r < 0$.

**Example 29.** *Define*

$$a_{nk} = \begin{cases} nC_k r^k (1-r)^{n-k}, & \text{if } 0 \le k \le n; \\ 0, & \text{if } k > n, \end{cases}$$

*where, as usual,*

$$nC_k = \frac{n!}{k!(n-k)!}.$$

*One can show that the above matrix is regular if* $0 < r < 1$. *It is called the Euler–Knopp matrix of order* $r$, *denoted by* $(E, r)$.

**Example 30.** *Natarajan introduced the* $(M, \lambda_n)$ *method as follows: Let* $\{\lambda_n\}$ *be a sequence such that* $\sum_{n=0}^{\infty} |\lambda_n| < \infty$. *The* $(M, \lambda_n)$ *method is defined by the infinite matrix* $(a_{nk})$, *where*

$$a_{nk} = \begin{cases} \lambda_{n-k}, & \text{if } k \le n; \\ 0, & \text{if } k > n. \end{cases}$$

*In this context, we note that the* $(M, \lambda_n)$ *method reduces to the well-known* $Y$*-method when*

$$\lambda_0 = \lambda_1 = \frac{1}{2} \text{ and } \lambda_n = 0, \text{ for all } n = 2, 3, \ldots.$$

*We can prove that the* $(M, \lambda_n)$ *method is regular if and only if*

$$\sum_{n=0}^{\infty} \lambda_n = 1. \tag{8.14}$$

**Example 31.** *The weighted mean method or* $(\overline{N}, p_n)$ *method is defined by the infinite matrix* $A = (a_{nk})$, *where*

$$a_{nk} = \begin{cases} \frac{p_k}{P_n}, & \text{if } k \le n; \\ 0, & \text{if } k > n, \end{cases}$$

$P_n = \sum_{k=0}^{n} p_k$, $P_n \neq 0$, $n = 0, 1, 2, \ldots$. *One can show that the weighted mean method* $(\overline{N}, p_n)$ *is regular if and only if*

$$\sum_{k=0}^{n} |p_k| = O(P_n), \; n \to \infty; \tag{8.15}$$

*and*

$$P_n \to \infty, \ n \to \infty. \tag{8.16}$$

**Example 32.** *The sequence* $\{s_n\}$ *is said to be Abel summable to* $s$ *if*

$$\lim_{x \to 1-} (1-x) \sum_{n=0}^{\infty} s_n x^n \text{ exists and } = s.$$

*We can prove that the Abel method is regular. In this context, it is worth noting that the Abel method cannot be described by an infinite matrix.*

**Example 33.** *Nörlund method or* $(N, p_n)$ *method. Let* $p_0 > 0$, $p_n \geq 0$, $n = 1, 2, \ldots$ *and* $P_n = \sum_{k=0}^{n} p_k$, $n = 0, 1, 2, \ldots$. *Then the Nörlund method or the* $(N, p_n)$ *method is defined by the matrix* $(a_{nk})$, *where*

$$a_{nk} = \begin{cases} \frac{p_{n-k}}{P_n}, & \text{if } k \leq n; \\ 0, & \text{if } k > n. \end{cases}$$

*We can prove that the Nörlund method is regular if and only if*

$$\lim_{n \to \infty} \frac{p_n}{P_n} = 0.$$

For a more detailed study of several methods of summability, the reader can refer to ([3, 8, 13, 21, 22]) listed in the Bibliography at the end of the book.

---

## 8.3 Schur's theorem

In this section, we prove another important result of summability theory, viz., Schur's theorem.

**Definition 81.** $A = (a_{nk})$ *is called a Schur matrix if* $A \in (\ell_\infty, c)$, *where, as usual,* $\ell_\infty$ *denotes the Banach space of all bounded sequences, i.e.,* $\{(Ax)_n\} \in c$, *whenever,* $x = \{x_k\} \in \ell_\infty$.

The following result gives a characterization of a Schur matrix in terms of the entries of the matrix.

**Theorem 84** (Schur). *$A = (a_{nk})$ is a Schur matrix if and only if (8.12) holds and*

$$\sum_{k=0}^{\infty} |a_{nk}| \text{ converges uniformly in } n. \tag{8.17}$$

*Proof.* Sufficiency part. Let (8.12) and (8.17) hold. Condition (8.17) implies that

$$\sum_{k=0}^{\infty} |a_{nk}| \text{ converges, for every } n = 0, 1, 2, \ldots.$$

Note that (8.12) and (8.17) imply that

$$\sup_{n\geq 0} \sum_{k=0}^{\infty} |a_{nk}| = M < \infty.$$

Thus, for each $r = 0, 1, 2, \ldots,$ we have,

$$\lim_{n\to\infty} \sum_{k=0}^{r} |a_{nk}| \leq M.$$

Thus,

$$\sum_{k=0}^{r} |\delta_k| \leq M, \text{ for every } r = 0, 1, 2, \ldots,$$

and so,

$$\sum_{k=0}^{\infty} |\delta_k| < \infty.$$

Hence, if $x = \{x_k\} \in \ell_\infty$, it follows that $\sum_{k=0}^{\infty} a_{nk}x_k$ converges absolutely and uniformly in $n$. Consequently,

$$\begin{aligned} \lim_{n\to\infty} (Ax)_n &= \lim_{n\to\infty} \sum_{k=0}^{\infty} a_{nk}x_k \\ &= \sum_{k=0}^{\infty} \delta_k x_k, \end{aligned}$$

proving that $\{(Ax)_n\} \in c$, i.e., $A \in (\ell_\infty, c)$, proving the sufficiency part.

Necessity part. Let $A = (a_{nk}) \in (\ell_\infty, c)$. So $A \in (c, c)$ and so (8.12) holds. Since $A \in (c, c)$, (8.1) holds, i.e.,

$$\sup_{n \geq 0} \sum_{k=0}^{\infty} |a_{nk}| < \infty.$$

As in the sufficiency part of the theorem, it follows that

$$\sum_{k=0}^{\infty} |\delta_k| < \infty.$$

We write

$$b_{nk} = a_{nk} - \delta_k, \ n, k = 0, 1, 2, \ldots.$$

Then

$$\left\{ \sum_{k=0}^{\infty} b_{nk} x_k \right\}_{n=0}^{\infty} \quad \text{converges, for all } x = \{x_k\} \in \ell_\infty.$$

We now claim that

$$\sum_{k=0}^{\infty} |b_{nk}| \to 0, \ n \to \infty. \tag{8.18}$$

Suppose not. Then

$$\overline{\lim_{n \to \infty}} \sum_{k=0}^{\infty} |b_{nk}| = h > 0.$$

So

$$\sum_{k=0}^{\infty} |b_{mk}| \to h, \ m \to \infty$$

through some subsequence of positive integers. We also note that

$$\lim_{m \to \infty} b_{mk} = 0, \ k = 0, 1, 2, \ldots.$$

We can now find a positive integer $m(1)$ such that

$$\left| \sum_{k=0}^{\infty} |b_{m(1),k}| - h \right| < \frac{h}{10}$$

and

$$|b_{m(1),0}| + |b_{m(1),1}| < \frac{h}{10}.$$

Since $\sum_{k=0}^{\infty} |b_{m(1),k}| < \infty$, we can choose $k(2) > 1$ such that

$$\sum_{k=k(2)+1}^{\infty} |b_{m(1),k}| < \frac{h}{10}.$$

It now follows that

$$\begin{aligned}\left|\sum_{k=2}^{k(2)} |b_{m(1),k}| - h\right| &= \left|\left(\sum_{k=0}^{\infty} |b_{m(1),k}| - h\right) - (|b_{m(1),0}| + |b_{m(1),1}|)\right. \\ &\qquad \left. - \sum_{k=k(2)+1}^{\infty} |b_{m(1),k}|\right| \\ &< \frac{h}{10} + \frac{h}{10} + \frac{h}{10} \\ &= \frac{3h}{10}.\end{aligned}$$

Now, choose a positive integer $m(2) > m(1)$ such that

$$\left|\sum_{k=0}^{\infty} |b_{m(2),k}| - h\right| < \frac{h}{10}$$

and

$$\sum_{k=0}^{k(2)} |b_{m(2),k}| < \frac{h}{10}.$$

Then choose a positive integer $k(3) > k(2)$ such that

$$\sum_{k=k(3)+1}^{\infty} |b_{m(2),k}| < \frac{h}{10}.$$

It now follows, as before, that

$$\left|\sum_{k=k(2)+1}^{k(3)} |b_{m(2),k}| - h\right| < \frac{3h}{10}.$$

Continuing in this manner, we find $m(1) < m(2) < \ldots$ and $1 = k(1) < k(2) < k(3) < \ldots$ such that

$$\sum_{k=0}^{k(r)} |b_{m(r),k}| < \frac{h}{10}, \tag{8.19}$$

$$\sum_{k=k(r+1)+1}^{\infty} |b_{m(r),k}| < \frac{h}{10}, \tag{8.20}$$

and

$$\left| \sum_{k=k(r)+1}^{k(r+1)} |b_{m(r),k}| - h \right| < \frac{3h}{10}. \tag{8.21}$$

Define the sequence $x = \{x_k\}$ as follows:

$$x_0 = x_1 = 0$$

and

$$x_k = (-1)^r sgn b_{m(r),k}, \text{ if } k(r) < k \leq k(r+1), \; r = 1, 2, \ldots.$$

Note that $x = \{x_k\} \in \ell_\infty$ and $\|x\| = 1$. Now,

$$\begin{aligned}
&\left| \sum_{k=0}^{\infty} b_{m(r),k} x_k - (-1)^r h \right| \\
&= \left| \sum_{k=0}^{k(r)} b_{m(r),k} x_k + \sum_{k=k(r)+1}^{k(r+1)} b_{m(r),k} x_k \right. \\
&\qquad \left. + \sum_{k=k(r+1)+1}^{\infty} b_{m(r),k} x_k - (-1)^r h \right| \\
&= \left| \left\{ \sum_{k=k(r)+1}^{k(r+1)} |b_{m(r),k}| - h \right\} (-1)^r \right. \\
&\qquad \left. + \sum_{k=0}^{k(r)} b_{m(r),k} x_k + \sum_{k=k(r+1)+1}^{\infty} b_{m(r),k} x_k \right| \\
&< \frac{3h}{10} + \frac{h}{10} + \frac{h}{10}, \text{ using (8.19), (8.20) and (8.21)} \\
&= \frac{h}{2}.
\end{aligned}$$

Consequently, $\left\{ \sum_{k=0}^{\infty} b_{nk} x_k \right\}_{n=0}^{\infty}$ is not a Cauchy sequence and so it is not convergent, which is a contradiction. Hence (8.18) holds. So, given $\epsilon > 0$, there exists a positive integer $n_0$ such that

$$\sum_{k=0}^{\infty} |b_{nk}| < \epsilon, \text{ for all } n > n_0. \tag{8.22}$$

Since

$$\sum_{k=0}^{\infty} |b_{nk}| < \infty, \text{ for } 0 \le n \le n_0,$$

we can find a positive integer $M$ such that

$$\sum_{k=M}^{\infty} |b_{nk}| < \epsilon, n = 0, 1, 2, \ldots, n_0. \tag{8.23}$$

In view of (8.22) and (8.23), we have

$$\sum_{k=M}^{\infty} |b_{nk}| < \epsilon, \text{ for all } n = 0, 1, 2, \ldots,$$

$$\textit{i.e.,} \sum_{k=0}^{\infty} |b_{nk}| \text{ converges uniformly in } n.$$

Since $\sum_{k=0}^{\infty} |\delta_k| < \infty$, it follows that $\sum_{k=0}^{\infty} |a_{nk}|$ converges uniformly in $n$, proving (8.17) and the necessity part. The proof of Schur's theorem is now complete.

□

**Remark 7.** *We can suitably modify proof of Theorem 84 to prove that if $A = (a_{nk})$ sums all sequences of 0's and 1's, then $A$ is a Schur matrix. The converse is trivially true. Thus we have Hahn's version of Schur's theorem.*

**Theorem 85** (Hahn). *$A$ is a Schur matrix if and only if $A$ sums all sequences of 0's and 1's.*

The author improved Hahn's theorem and proved the following result.

**Theorem 86** (Natarajan). *$A$ is a Schur matrix if and only if $A$ sums all non-periodic sequences of 0's and 1's.*

---

## 8.4 Steinhaus theorem and its improvement

Using Theorem 82 and Theorem 84, we can deduce the following very important result due to Steinhaus, which can be proved independently too.

**Theorem 87** (Steinhaus). *An infinite matrix cannot be both a regular and a Schur matrix. In other words, given a regular matrix, there exists a bounded divergent sequence that is not A-summable. Writing in symbols, the theorem means that*

$$(c, c; P) \cap (\ell_\infty, c) = \phi.$$

*Proof.* Suppose $A = (a_{nk})$ is both a regular and a Schur matrix. Then (8.2) and (8.3) hold. In view of (8.17),

$$\begin{aligned}\lim_{n\to\infty} \sum_{k=0}^{\infty} a_{nk} &= \sum_{k=0}^{\infty} \lim_{n\to\infty} a_{nk} \\ &= 0, \text{ using (8.2)},\end{aligned}$$

which contradicts (8.3), proving the theorem. □

In the context of Steinhaus theorem, the author introduced the sequence space $\Lambda_r$, $r \geq 1$ being a fixed integer, as follows:

**Definition 82.** *$\Lambda_r$ is defined as the set of all sequences $x = \{x_k\} \in \ell_\infty$ such that*

$$|x_{k+r} - x_k| \to 0, \ k \to \infty,$$

*$r \geq 1$ being a fixed integer.*

It is easily seen that $\Lambda_r$ is a closed subspace of $\ell_\infty$ under the norm $\|x\| = \sup_{k\geq 0} |x_k|$, $x = \{x_k\} \in \Lambda_r$.

The following result, improving the Steinhaus theorem, was proved by the author. We give a constructive proof.

**Theorem 88** (Natarajan).

$$(c, c; P) \cap \left(\Lambda_r - \bigcup_{i=1}^{r-1} \Lambda_i, c\right) = \phi.$$

*Proof.* Let $A$ be a regular matrix, i.e., $A \in (c, c; P)$. We can now choose two sequences $\{n(m)\}$, $\{k(m)\}$ of positive integers such that if

$$m = 2p, \ n(m) > n(m-1), \ k(m) > k(m-1) + (2m-5)r,$$

then

$$\sum_{k=0}^{k(m-1)+(2m-5)r} |a_{n(m),k}| < \frac{1}{16},$$

$$\sum_{k=k(m-1)}^{\infty} |a_{n(m),k}| < \frac{1}{16};$$

and if

$$m = 2p + 1, \; n(m) > n(m-1), \; k(m) > k(m-1) + (m-2)r,$$

then

$$\sum_{k=0}^{k(m-1)+(m-2)r} |a_{n(m),k}| < \frac{1}{16},$$

$$\sum_{k=k(m-1)+(m-2)r+1}^{k(m)} |a_{n(m),k}| > \frac{7}{8},$$

$$\sum_{k=k(m)+1}^{\infty} |a_{n(m),k}| < \frac{1}{16}.$$

Define the sequence $x = \{x_k\}$ as follows:

if $k(2p-1) < k \le k(2p)$, then,

$$x_k = \begin{cases} \frac{2p-2}{2p-1}, & \text{if } k = k(2p-1)+1; \\ 1, & \text{if } k(2p-1)+1 < k \le k(2p-1)+r; \\ \frac{2p-3}{2p-1}, & \text{if } k = k(2p-1)+r+1; \\ 1, & \text{if } k(2p-1)+r+1 < k \le k(2p-1)+2r; \\ \vdots & \\ 1, & \text{if } k(2p-1)+(2p-4)r+1 < k \le k(2p-1)+(2p-3)r; \\ \frac{1}{2p-1}, & \text{if } k = k(2p-1)+(2p-3)r+1; \\ \frac{2p-2}{2p-1}, & \text{if } k(2p-1)+(2p-3)r+1 < k \le k(2p-1)+(2p-2)r; \\ 0, & \text{if } k = k(2p-1)+(2p-2)r+1; \\ \vdots & \\ \frac{1}{2p-1}, & \text{if } k(2p-1)+(4p-6)r+1 < k \le k(2p-1)+(4p-5)r; \\ 0, & \text{if } k(2p-1)+(4p-5)r < k \le k(2p), \end{cases}$$

and if $k(2p) < k \le k(2p+1)$, then

$$x_k = \begin{cases} \frac{1}{2p}, & \text{if } k(2p) < k \le k(2p)+r; \\ \frac{2}{2p}, & \text{if } k(2p)+r < k \le k(2p)+2r; \\ \vdots & \\ \frac{2p-1}{2p}, & \text{if } k(2p)+(2p-2)r < k \le k(2p)+(2p-1)r; \\ 1, & \text{if } k(2p)+(2p-1)r < k \le k(2p+1). \end{cases}$$

Note that

$$\text{if } k(2p-1) < k \le k(2p),$$
$$|x_{k+r} - x_k| < \tfrac{1}{2p-1},$$

and

$$\text{if } k(2p) < k \le k(2p+1),$$
$$|x_{k+r} - x_k| < \tfrac{1}{2p}.$$

Thus,

$$|x_{k+r} - x_k| \to 0, \ k \to \infty$$

and so $x = \{x_k\} \in \Lambda_r$. However,

$$|x_{k+1} - x_k| = \frac{2p-2}{2p-1}, \text{ if } k = k(2p-1) + (2p-3)r, \ p = 1, 2, \ldots,$$

so that

$$|x_{k+1} - x_k| \not\to 0, \ k \to \infty$$

and consequently,

$$x = \{x_k\} \notin \Lambda_1.$$

In a similar fashion, we can prove that

$$x = \{x_k\} \notin \Lambda_i, \ i = 2, 3, \ldots, (r-1).$$

Thus,

$$x \notin \bigcup_{i=1}^{r-1} \Lambda_i$$

and so

$$x \in \Lambda_r - \bigcup_{i=1}^{r-1} \Lambda_i.$$

However,

$$\left.\begin{aligned} |(Ax)_{n(2p)}| &< \tfrac{1}{16} + \tfrac{1}{16} = \tfrac{1}{8}, \\ |(Ax)_{n(2p+1)}| &> \tfrac{7}{8} - \tfrac{1}{16} - \tfrac{1}{16} = \tfrac{3}{4} \end{aligned}\right\}, \ p = 1, 2, \ldots,$$

proving that $\{(Ax)_n\} \notin c$, completing the proof of the theorem. □

**Remark 8.** *We note that*

$$(c, c; P) \cap (\Lambda_r, c) = \phi, \ r \text{ being an integer } \geq 1.$$

*Since*

$$(\ell_\infty, c) \subset (\Lambda_r, c),$$

*it follows that*

$$(c, c; P) \cap (\ell_\infty, c) = \phi,$$

*which is the Steinhaus theorem, thus getting an improvement of the Steinhaus theorem.*

The sequence spaces $\Lambda_r$, $r \geq 1$ being an integer, have an important role to play in the context of sequences of 0's and 1's and the reader can refer to [20] for a detailed study.

**Exercises**

1. Use the Banach–Steinhaus theorem to prove the Silverman–Toeplitz theorem.

2. Let
$$\lim_{n\to\infty} a_{nk} = 0,\ k = 0, 1, 2, \ldots$$
and
$$M = \sup_{n\geq 0} \sum_{k=0}^{\infty} |a_{nk}| < \infty.$$
Prove that $A = (a_{nk})$ defines a bounded linear mapping of $c_0$ into itself and $\|A\| = M$.

3. Let $A$ be a bounded linear mapping of $c_0$ into itself. Prove, using the Banach–Steinhaus theorem, that $A$ determines an infinite matrix $(a_{nk})$ such that
$$(Ax)_n = \sum_{k=0}^{\infty} a_{nk}x_k,\ \text{for every } x = \{x_k\} \in c_0,$$
$$\lim_{n\to\infty} a_{nk} = 0,\ k = 0, 1, 2, \ldots$$
and
$$\|A\| = \sup_{n\geq 0} \sum_{k=0}^{\infty} |a_{nk}| < \infty.$$

4. Characterize the matrix classes $(\ell_\infty, \ell_\infty)$ and $(c_0, c_0)$.

5. Characterize the matrix classes $(c_0, c)$ and $(c, c_0)$.

6. Prove that $A = (a_{nk}) \in (\ell_\infty, c_0)$ if and only if
$$\sum_{k=0}^{\infty} |a_{nk}| \to 0,\ n \to \infty.$$

7. If $A = (a_{nk})$, $B = (b_{nk}) \in (c, c)$, prove that $A + B$, $AB \in (c, c)$, where $AB$ is the usual matrix product, i.e.,
$$(AB)_{nk} = \sum_{i=0}^{\infty} a_{ni}b_{ik},\ n, k = 0, 1, 2, \ldots.$$

8. Find the necessary and sufficient conditions for $A = (a_{nk}) \in (\ell_1, \ell_1)$.

9. Let $1 < p < \infty$ and $A = (a_{nk}) \in (\ell_\infty, \ell_\infty) \cap (\ell_1, \ell_1)$. Prove that $A \in (\ell_p, \ell_p)$.

10. If $\sum_{k=0}^{\infty} |a_{nk}| < \infty$, for every $n = 0, 1, 2, \dots$ and

$$\sum_{k=0}^{\infty} |a_{nk}| \to 0, \ n \to \infty,$$

prove that $\sum_{k=0}^{\infty} |a_{nk}|$ is uniformly convergent in $n$.

11. Prove that $A$ is a Schur matrix if and only if $A$ sums all sequences of 0's and 1's.

12. Prove the Steinhaus theorem without using Schur's theorem, i.e., given a regular matrix, construct a bounded divergent sequence $x = \{x_k\}$ such that $\{(Ax)_n\}$ diverges.

13. Characterize the matrix class $(\ell_1, \ell_\infty)$.

14. Find the necessary and sufficient conditions for $A = (a_{nk}) \in (\ell_1, c)$.

15. Determine if $(\ell_1, \ell_p)$, $1 \leq p \leq \infty$, is a Banach space under the norm

$$\|A\| = \sup_{k \geq 0} \left( \sum_{n=0}^{\infty} |a_{nk}|^p \right)^{\frac{1}{p}}.$$

Justify your answer.

16. Characterize the matrix classes $(c, \ell_1)$ and $(\ell_\infty, \ell_1)$.

17. Prove that $A = (a_{nk}) \in (\ell_1, \ell_p)$, $1 \leq p < \infty$, if and only if

$$\sup_{k \geq 0} \sum_{n=0}^{\infty} |a_{nk}|^p < \infty.$$

18. We write $A = (a_{nk}) \in (\ell_1, \ell_1; P)$ if $A \in (\ell_1, \ell_1)$ and $\sum_{n=0}^{\infty} (Ax)_n = \sum_{k=0}^{\infty} x_k$, $x = \{x_k\} \in \ell_1$. Find the necessary and sufficient conditions for $A \in (\ell_1, \ell_1; P)$.

# Chapter 9

# Steinhaus-Type Theorems

## 9.1 Some Steinhaus-type theorems

Whenever there is some notion of limit or sum in the sequence spaces $X, Y$, we denote by $(X, Y; P)$ that subclass of $(X, Y)$ consisting of all infinite matrices, which preserve this limit or sum. Then results of the form

$$(X, Y; P) \cap (Z, Y) = \phi,$$

$X \subsetneqq Z$, are called "Steinhaus-type" theorems. When the limit in $X$ and the sum in $Y$ are preserved or vice versa, then the class of all such matrices is denoted by $(X, Y; P')$. Note the difference between the classes $(X, Y; P)$ and $(X, Y; P')$. Then results of the form

$$(X, Y; P') \cap (Z, Y) = \phi,$$

$X \subsetneqq Z$, are also called "Steinhaus-type" theorems.

In the present section, we prove some Steinhaus-type theorems.

The sequence spaces $\ell_p$, $p \geq 1$, $c_0, c, \ell_\infty$ are defined as usual:

$$\ell_p = \left\{ x = \{x_k\} / \sum_{k=0}^{\infty} |x_k|^p < \infty \right\}, \; p \geq 1,$$

$$c_0 = \left\{ x = \{x_k\} / \lim_{k\to\infty} x_k = 0 \right\},$$

$$c = \left\{ x = \{x_k\} / \lim_{k\to\infty} x_k \text{ exists} \right\},$$

$$\ell_\infty = \left\{ x = \{x_k\} / \sup_{k\geq 0} |x_k| < \infty \right\}.$$

Note that

$$\ell_p \subset c_0 \subset c \subset \ell_\infty, \text{ where } p \geq 1.$$

In this chapter, for convenience, we write $\ell_1 = \ell$. Let $(\ell, c; P')$ denote the class of all infinite matrices $A \in (\ell, c)$ such that

$$\lim_{n\to\infty} (Ax)_n = \sum_{k=0}^{\infty} x_k, \; x = \{x_k\} \in \ell.$$

We now state the following result (see Exercise 11 in Chapter 8).

**Theorem 89.** *$A = (a_{nk}) \in (\ell, c)$ if and only if*

$$\sup_{n,k} |a_{nk}| < \infty \tag{9.1}$$

*and (8.12) holds.*

The following result can now be proved.

**Theorem 90.** *$A = (a_{nk}) \in (\ell, c; P')$ if and only if (9.1), (8.12) hold with $\delta_k = 1$, $k = 0, 1, 2, \ldots$.*

*Proof.* Let $A \in (\ell, c; P')$. Then (9.1) holds. Let $e_k$ be the sequence in which 1 occurs in the $k$th place and 0 elsewhere, i.e.,

$$e_k = \{x_i^{(k)}\}_{i=0}^{\infty},$$

where

$$x_i^{(k)} = \begin{cases} 1, & \text{if } i = k; \\ 0, & \text{if } i \neq k, \; k = 0, 1, 2, \ldots. \end{cases}$$

Then $e_k \in \ell$, $k = 0, 1, 2, \ldots$ and $\sum_{i=0}^{\infty} x_i^{(k)} = 1$, $k = 0, 1, 2, \ldots$. Now, $(Ae_k)_n = a_{nk}$ so that

$$\lim_{n \to \infty} a_{nk} = 1, \ k = 0, 1, 2, \ldots,$$

$$i.e., \ \delta_k = 1, \ k = 0, 1, 2, \ldots.$$

Conversely, let (9.1) hold and (8.12) hold with $\delta_k = 1$, $k = 0, 1, 2, \ldots$. Let $x = \{x_k\} \in \ell$. In view of (9.1), $(Ax)_n$ is defined, for $n = 0, 1, 2, \ldots$. Now,

$$\begin{aligned}(Ax)_n &= \sum_{k=0}^{\infty} a_{nk} x_k \\ &= \sum_{k=0}^{\infty} (a_{nk} - 1) x_k + \sum_{k=0}^{\infty} x_k,\end{aligned}$$

this being true since both the series on the right-hand side converge. Since $\sum_{k=0}^{\infty} |x_k| < \infty$, given $\epsilon > 0$, there exists a positive integer $N$ such that

$$\sum_{k=N+1}^{\infty} |x_k| < \frac{\epsilon}{2A}, \tag{9.2}$$

where $A = \sup_{n,k} |a_{nk} - 1|$. Since,

$$\lim_{n \to \infty} a_{nk} = 1, \ k = 0, 1, 2, \ldots, N,$$

we can choose a positive integer $N' > N$ such that

$$|a_{nk} - 1| < \frac{\epsilon}{2(N+1)M}, \text{ for all } n \geq N', \ k = 0, 1, 2, \ldots, N, \tag{9.3}$$

where $M > 0$ is such that $|x_k| \leq M$, $k = 0, 1, 2, \ldots$. Now, for $n \geq N'$,

$$\begin{aligned}\left| \sum_{k=0}^{\infty} (a_{nk} - 1) x_k \right| &\leq \sum_{k=0}^{N} |a_{nk} - 1||x_k| \\ &\quad + \sum_{k=N+1}^{\infty} |a_{nk} - 1||x_k| \\ &< (N+1) \frac{\epsilon}{2(N+1)M} M + A \frac{\epsilon}{2A}, \\ &\quad \text{using (9.2) and (9.3)} \\ &= \epsilon,\end{aligned}$$

so that

$$\lim_{n\to\infty}\sum_{k=0}^{\infty}(a_{nk}-1)x_k = 0.$$

Consequently,

$$\lim_{n\to\infty}(Ax)_n = \sum_{k=0}^{\infty} x_k,$$

and so $A \in (\ell, c; P')$, completing the proof of the theorem. □

Using Theorem 90, we have the following Steinhaus-type result.

**Theorem 91.** $(\ell, c; P') \cap (\ell_p, c) = \phi$, $p > 1$.

*Proof.* Let $A = (a_{nk}) \in (\ell, c; P') \cap (\ell_p, c)$, $p > 1$. It is known that $A \in (\ell_p, c)$, $p > 1$, if and only if (8.12) holds and

$$\sup_{n\geq 0}\sum_{k=0}^{\infty}|a_{nk}|^q < \infty, \tag{9.4}$$

where $\frac{1}{p}+\frac{1}{q}=1$. It now follows that

$$\sum_{k=0}^{\infty}|\delta_k|^q < \infty,$$

which contradicts the fact that $\delta_k = 1$, $k = 0, 1, 2, \ldots$, since $A \in (\ell, c; P')$ and consequently, $\sum_{k=0}^{\infty}|\delta_k|^q$ diverges, proving the theorem. □

**Corollary 9.** *Since* $\ell_p \subset c_0 \subset c \subset \ell_\infty$, $p > 1$,

$$(\ell_\infty, c) \subset (c, c) \subset (c_0, c) \subset (\ell_p, c),\ \ p > 1,$$

*so that*

$$(\ell, c; P') \cap (X, c) = \phi,$$

*when*

$$X = \ell_\infty, c, c_0, \ell_p,\ \ p > 1.$$

**Remark 9.** *We recall that $\ell_\infty$ is a Banach space with respect to the norm $\|x\| = \sup_{k\geq 0} |x_k|$, $x = \{x_k\} \in \ell_\infty$ and $c_0$, $c$ are closed subspaces of $\ell_\infty$. $\ell_p$, $p > 1$ is a Banach space under the norm*

$$\|x\| = \left(\sum_{k=0}^{\infty} |x_k|^p\right)^{\frac{1}{p}}, \quad x = \{x_k\} \in \ell_p.$$

*Let, as usual, $B(\ell, c)$ denote the space of all bounded linear mappings of $\ell$ into $c$. We note that if $A = (a_{nk}) \in (\ell, c; P')$, then $A$ is bounded and*

$$\|A\| = \sup_{n,k} |a_{nk}|.$$

*However, $(\ell, c; P')$ is not a subspace of $B(\ell, c)$, since, if $A \in (\ell, c; P')$, $\lim_{n\to\infty} 2a_{nk} = 2$, $k = 0, 1, 2, \ldots$ and so $2A \notin (\ell, c; P')$.*

We now need the following sequence space:

Let

$$\gamma = \left\{x = \{x_k\} / \sum_{k=0}^{\infty} x_k \text{ converges}\right\}.$$

We note that $\gamma \subset c_0 \subset c \subset \ell_\infty$.

We now write $A \in (\ell, \ell; P)$ if $A \in (\ell, \ell)$ and

$$\sum_{n=0}^{\infty} (Ax)_n = \sum_{k=0}^{\infty} x_k, \ x = \{x_k\} \in \ell.$$

We also write

$$A \in (\ell, \ell; P)'$$

if $A \in (\ell, \ell; P)$ and

$$\lim_{k\to\infty} a_{nk} = 0, \ n = 0, 1, 2, \ldots. \tag{9.5}$$

One can prove that $A \in (\ell, \ell; P)$ if and only if $A \in (\ell, \ell)$ and

$$\sum_{n=0}^{\infty} a_{nk} = 1, \ k = 0, 1, 2, \ldots. \tag{9.6}$$

Maddox [12] noted that

$$(\ell, \ell; P) \cap (\gamma, \ell) \neq \phi.$$

We now prove that given a $(\ell, \ell; P)'$ matrix $A$, there exists a sequence $x = \{x_k\} \in \gamma$, whose $A$-transform is not in $\ell$. In particular, a lower triangular $(\ell, \ell; P)$ matrix cannot belong to the class $(\gamma, \ell)$. We use the following lemma, the proof of which is modeled on Fridy's result (see [6]).

**Lemma 6.** *If $A = (a_{nk}) \in (\ell, \ell)$ and satisfies (9.5), and if*

$$\overline{\lim_{k\to\infty}} \left|\sum_{n=0}^{\infty} a_{nk}\right| > 0, \tag{9.7}$$

*then there exists a sequence $x = \{x_k\} \in \gamma$ such that $\{(Ax)_n\} \notin \ell$.*

*Proof.* By hypothesis, for some $\epsilon > 0$, there exists a strictly increasing sequence $\{k(i)\}$ of positive integers such that

$$\left|\sum_{n=0}^{\infty} a_{n,k(i)}\right| \geq 2\epsilon, \; i = 1, 2, \ldots.$$

In particular,

$$\left|\sum_{n=0}^{\infty} a_{n,k(1)}\right| \geq 2\epsilon.$$

We then choose a positive integer $n(1)$ such that

$$\sum_{n>n(1)} |a_{n,k(1)}| < \min(\frac{1}{2}, \frac{\epsilon}{2}),$$

which is possible because

$$\sum_{n=0}^{\infty} |a_{nk}| < \infty, \; k = 0, 1, 2, \ldots,$$

in view of the fact that $A \in (\ell, \ell)$. Now it follows that

$$\left|\sum_{n=0}^{n(1)} a_{n,k(1)}\right| > \epsilon.$$

In general, having chosen $k(j), n(j), \; j \leq m-1$, choose a positive integer $k(m) > k(m-1)$ such that

$$\left|\sum_{n=0}^{\infty} a_{n,k(m)}\right| \geq 2\epsilon,$$

$$\sum_{n=0}^{n(m-1)} |a_{n,k(m)}| < \min(\frac{1}{2}, \frac{\epsilon}{2}),$$

and then choose a positive integer $n(m) > n(m-1)$ such that

$$\sum_{n>n(m)} |a_{n,k(m)}| < \min(\frac{1}{2^m}, \frac{\epsilon}{2}),$$

so that

$$\left| \sum_{n=n(m-1)+1}^{n(m)} a_{n,k(m)} \right| > 2\epsilon - \frac{\epsilon}{2} - \frac{\epsilon}{2}$$
$$= \epsilon.$$

Consider the sequence $x = \{x_k\}$ defined by:

$$x_k = \begin{cases} \frac{(-1)^{i+1}}{i}, & \text{if } k = k(i); \\ 0, & \text{if } k \neq k(i), i = 1, 2, \ldots. \end{cases}$$

It is clear that $x = \{x_k\} \in \gamma$. Defining $n(0) = 0$, we have

$$\begin{aligned}
\sum_{n=0}^{n(N)} |(Ax)_n| &\geq \sum_{m=1}^{N} \sum_{n=n(m-1)+1}^{n(m)} |(Ax)_n| \\
&= \sum_{m=1}^{N} \sum_{n=n(m-1)+1}^{n(m)} \left| \sum_{i=1}^{\infty} a_{n,k(i)} x_{k(i)} \right| \\
&= \sum_{m=1}^{N} \sum_{n=n(m-1)+1}^{n(m)} \left| \sum_{i=1}^{\infty} a_{n,k(i)} \frac{(-1)^{i+1}}{i} \right| \\
&\geq \sum_{m=1}^{N} \sum_{n=n(m-1)+1}^{n(m)} \left\{ \left| \frac{(-1)^{m+1}}{m} a_{n,k(m)} \right| \right. \\
&\qquad \left. - \sum_{\substack{i=1 \\ i \neq m}}^{\infty} \left| \frac{(-1)^{i+1}}{i} a_{n,k(i)} \right| \right\} \\
&= \sum_{m=1}^{N} \sum_{n=n(m-1)+1}^{n(m)} \left\{ \frac{1}{m} |a_{n,k(m)}| \right. \\
&\qquad \left. - \sum_{\substack{i=1 \\ i \neq m}}^{\infty} \frac{1}{i} |a_{n,k(i)}| \right\} \\
&> \epsilon \sum_{m=1}^{N} \frac{1}{m} - \sum_{m=1}^{N} \sum_{n=n(m-1)+1}^{n(m)} \sum_{\substack{i=1 \\ i \neq m}}^{\infty} |a_{n,k(i)}|, \\
&\qquad \text{since } \frac{1}{i} \leq 1.
\end{aligned}$$

Now

$$\begin{aligned}
\sum_{m=1}^{\infty} \sum_{n=n(m-1)+1}^{n(m)} \sum_{i<m} |a_{n,k(i)}| \\
&= \sum_{j=1}^{\infty} \sum_{i>m(j)} |a_{i,k(j)}| \\
&\leq \sum_{m=1}^{\infty} \frac{1}{2^m} \\
&= 1.
\end{aligned}$$

Similarly,

$$\sum_{m=1}^{\infty} \sum_{n=n(m-1)+1}^{n(m)} \sum_{i>m} |a_{n,k(i)}| \leq \frac{1}{2}.$$

Consequently,

$$\sum_{n=0}^{n(N)} |(Ax)_n| > \epsilon \sum_{m=1}^{N} \frac{1}{m} - \frac{3}{2}.$$

Since $\sum_{m=1}^{\infty} \frac{1}{m}$ diverges, it follows that $\{(Ax)_n\} \notin \ell$, completing the proof of the lemma. □

We shall also call the following a Steinhaus-type result.

**Theorem 92.** $(\ell, \ell; P)' \cap (\gamma, \ell) = \phi$.

*Proof.* Let $A = (a_{nk}) \in (\ell, \ell; P)' \cap (\gamma, \ell)$. Since (9.6) holds,

$$\overline{\lim_{k \to \infty}} \left| \sum_{n=0}^{\infty} a_{nk} \right| = 1 > 0.$$

In view of the preceding Lemma 6, there exists a sequence $x = \{x_k\} \in \gamma$ such that $\{(Ax)_n\} \notin \ell$, which is a contradiction, completing the proof. □

**Remark 10.** *If the condition (9.5) is dropped, the above theorem fails to hold, as the following example illustrates. The infinite matrix*

$$A = (a_{nk}) = \begin{pmatrix} 1 & 1 & 1 & \cdots \\ 0 & 0 & 0 & \cdots \\ 0 & 0 & 0 & \cdots \\ \cdots & \cdots & \cdots & \cdots \end{pmatrix}$$

*is in* $(\ell, \ell; P)' \cap (\gamma, \ell)$ *but* $a_{0k} \not\to 0$, $k \to \infty$.

In this context, Fridy [5] proved the following result.

**Theorem 93** (Fridy). $(\ell, \ell; P) \cap (\ell_p, \ell) = \phi$, *for* $p > 1$.

## 9.2 More Steinhaus-type theorems

We now introduce the sequence space

$$\gamma_\infty = \left\{ x = \{x_k\}/\{s_k\} \in \ell_\infty,\ s_k = \sum_{i=0}^{k} x_i,\ k = 0, 1, 2, \ldots \right\}.$$

We note that $\gamma_\infty \subset \ell_\infty$. $(\ell, \gamma; P)$ denotes the set of all infinite matrices $A = (a_{nk}) \in (\ell, \gamma)$ such that

$$\sum_{n=0}^{\infty} (Ax)_n = \sum_{k=0}^{\infty} x_k,\ x = \{x_k\} \in \ell.$$

The following result is known [25].

**Theorem 94.** *$A = (a_{nk}) \in (\ell, \gamma)$ if and only if*

$$\sup_{m,k} \left| \sum_{n=0}^{m} a_{nk} \right| < \infty; \tag{9.8}$$

*and*

$$\sum_{n=0}^{\infty} a_{nk} \text{ converges, for every } k = 0, 1, 2, \ldots. \tag{9.9}$$

We now have:

**Theorem 95.** *A matrix $A = (a_{nk}) \in (\ell, \gamma; P)$ if and only if (9.8), (9.9) hold and*

$$\sum_{n=0}^{\infty} a_{nk} = 1,\ \text{for every } k = 0, 1, 2, \ldots. \tag{9.10}$$

*Proof.* If $A \in (\ell, \gamma, P)$, then (9.8) holds. For $k = 0, 1, 2, \ldots$, let $e_k = \{0, 0, \ldots, 0, 1, 0, \ldots\}$, 1 occurring in the $k$th place. Then $e_k \in \ell$, $k = 0, 1, 2, \ldots$ and

$$\sum_{n=0}^{\infty} (Ae_k)_n = 1,\ k = 0, 1, 2, \ldots,$$

$$\text{i.e.,}\ \sum_{n=0}^{\infty} a_{nk} \text{ converges and } \sum_{n=0}^{\infty} a_{nk} = 1,\ k = 0, 1, 2, \ldots,$$

i.e., (9.9) and (9.10) hold.

Conversely, let (9.8), (9.9), and (9.10) hold. It is clear that $A \in (\ell, \gamma)$. Now, let $B = (b_{mk})$, where

$$b_{mk} = \sum_{n=0}^{m} a_{nk}, \; m, k = 0, 1, 2, \ldots .$$

Using (9.8) and (9.10) we have

$$\sup_{m,k} |b_{mk}| < \infty$$

and

$$\lim_{m \to \infty} b_{mk} = 1, \; k = 0, 1, 2, \ldots .$$

Now, using Theorem 90,

$$B \in (\ell, c; P').$$

Let $x = \{x_k\} \in \ell$. So,

$$\lim_{m \to \infty} \sum_{k=0}^{\infty} b_{mk} x_k \text{ exists and is equal to } \sum_{k=0}^{\infty} x_k,$$

$$\text{i.e., } \lim_{m \to \infty} \sum_{k=0}^{\infty} \left( \sum_{n=0}^{m} a_{nk} \right) x_k = \sum_{k=0}^{\infty} x_k,$$

$$\text{i.e., } \lim_{m \to \infty} \sum_{n=0}^{m} \left( \sum_{k=0}^{\infty} a_{nk} x_k \right) = \sum_{k=0}^{\infty} x_k,$$

$$\text{i.e., } \sum_{n=0}^{\infty} \left( \sum_{k=0}^{\infty} a_{nk} x_k \right) = \sum_{k=0}^{\infty} x_k,$$

$$\text{i.e., } \sum_{n=0}^{\infty} (Ax)_n = \sum_{k=0}^{\infty} x_k,$$

$$\text{i.e., } A \in (\ell, \gamma; P),$$

completing the proof of the theorem. □

The following result was proved by Maddox [12].

**Theorem 96** (Maddox). *$(\gamma,\gamma;P)\cap(\gamma_\infty,\gamma)=\phi$, where $(\gamma,\gamma;P)$ denotes the set of all matrices in $(\gamma,\gamma)$ such that*

$$\sum_{n=0}^{\infty}(Ax)_n=\sum_{k=0}^{\infty}x_k,\ x=\{x_k\}\in\gamma.$$

In the context of the above theorem, it is worthwhile to note that the identity matrix, i.e., $I=(i_{nk})$, where

$$i_{nk}=\begin{cases}1, & \text{if } k=n;\\ 0, & \text{if } k\neq n,\end{cases}$$

is in $(\ell,\gamma;P)\cap(\gamma_\infty,\gamma)$ so that

$$(\ell,\gamma;P)\cap(\gamma_\infty,\gamma)\neq\phi.$$

We have

$$(\gamma,\gamma;P)\subset(\ell,\gamma;P)$$

and

$$(c_0,\gamma)\subset(\gamma,\gamma).$$

"Enlarging" the class $(\gamma,\gamma;P)$ to $(\ell,\gamma;P)$, we would like to "contract" the class $(\gamma,\gamma)$ to $(c_0,\gamma)$ and attempt a Steinhaus-type result involving the classes $(\ell,\gamma;P)$ and $(c_0,\gamma)$.

**Theorem 97.** $(\ell,\gamma;P)\cap(c_0,\gamma)=\phi$.

*Proof.* Let $A=(a_{nk})\in(\ell,\gamma;P)\cap(c_0,\gamma)$. Since $A\in(c_0,\gamma)$,

$$\sup_{m\geq 0}\sum_{k=0}^{\infty}\left|\sum_{n=0}^{m}a_{nk}\right|\leq M<\infty,$$

for some $M>0$ (see [25]). Now, for $u=0,1,2,\ldots$,

$$\sum_{k=0}^{u}\left|\sum_{n=0}^{m}a_{nk}\right|\leq\sum_{k=0}^{\infty}\left|\sum_{n=0}^{m}a_{nk}\right|$$
$$\leq M,\ \text{for every } m=0,1,2,\ldots.$$

Taking the limit as $m\to\infty$, we have

$$\sum_{k=0}^{u}\left|\sum_{n=0}^{\infty}a_{nk}\right|\leq M,\ \text{for every } u=0,1,2,\ldots.$$

Now, taking the limit as $u \to \infty$, we get

$$\sum_{k=0}^{\infty}\left|\sum_{n=0}^{\infty} a_{nk}\right| \leq M,$$

which is a contradiction, since $\sum_{n=0}^{\infty} a_{nk} = 1$ for every $k = 0, 1, 2, \ldots$, in view of (9.10), establishing our claim. □

**Corollary 10.** *Since $c_0 \subset c \subset \ell_\infty$,*

$$(\ell_\infty, \gamma) \subset (c, \gamma) \subset (c_0, \gamma)$$

*so that*

$$(\ell, \gamma; P) \cap (X, \gamma) = \phi,$$

*when $X = c_0, c, \ell_\infty$.*

The following results are well known.

**Theorem 98.** *$A = (a_{nk}) \in (c_0, c)$ if and only if (8.1) and (8.12) hold (see Exercise 2 of Chapter 8).*

**Theorem 99.** *[25] $A = (a_{nk}) \in (\gamma, c; P')$ if and only if*

$$\sup_{n \geq 0}\left(\sum_{k=0}^{\infty} |a_{nk} - a_{n,k+1}|\right) < \infty \tag{9.11}$$

*and (8.12) holds with $\delta_k = 1$, $k = 0, 1, 2, \ldots$, where $(\gamma, c; P')$ is the subclass of $(\gamma, c)$ with*

$$\lim_{n\to\infty} (Ax)_n = \sum_{k=0}^{\infty} x_k, \;\; x = \{x_k\} \in \gamma.$$

A Steinhaus-type result now follows.

**Theorem 100.** $(\gamma, c; P') \cap (c_0, c) = \phi$.

*Proof.* Let $A = (a_{nk}) \in (\gamma, c; P') \cap (c_0, c)$. Since $A \in (c_0, c)$, using (8.1), it follows that

$$\sum_{k=0}^{\infty} |\delta_k| \text{ converges,}$$

which is a contradiction, since $\delta_k = 1$, $k = 0, 1, 2, \ldots$, in view of Theorem 99, completing the proof. □

**Corollary 11.** *Since* $c_0 \subset c \subset \ell_\infty$, $(\ell_\infty, c) \subset (c, c) \subset (c_0, c)$ *so that*

$$(\gamma, c; P') \cap (X, c) = \phi,$$

*when* $X = c_0, c, \ell_\infty$.

The following results are known [25], the proof of which is left to the reader.

**Theorem 101.** $A = (a_{nk}) \in (c_0, \gamma)$ *if and only if (9.9) holds and*

$$\sup_{m \geq 0} \sum_{k=0}^{\infty} \left| \sum_{n=0}^{m} a_{nk} \right| < \infty. \tag{9.12}$$

**Theorem 102.** $A = (a_{nk}) \in (\gamma, \gamma)$ *if and only if (9.9) holds and*

$$\sup_{m \geq 0} \sum_{k=0}^{\infty} \left| \sum_{n=0}^{m} (a_{nk} - a_{n,k-1}) \right| < \infty. \tag{9.13}$$

Further, $A \in (\gamma, \gamma, P)$ if and only if (9.10) and (9.13) hold, where we recall that $(\gamma, \gamma; P)$ is the subclass of $(\gamma, \gamma)$ such that

$$\sum_{n=0}^{\infty} (Ax)_n = \sum_{k=0}^{\infty} x_k, \quad x = \{x_k\} \in \gamma.$$

Consequently, we have the following Steinhaus-type result.

**Theorem 103.** $(\gamma, \gamma; P) \cap (c_0, \gamma) = \phi$.

*Proof.* Let $A = (a_{nk}) \in (\gamma, \gamma, P) \cap (c_0, \gamma)$. Using (9.12), it follows that

$$\sum_{k=0}^{\infty} \left| \sum_{n=0}^{\infty} a_{nk} \right| < \infty,$$

which is a contradiction, since $\sum_{n=0}^{\infty} a_{nk} = 1$, for every $k = 0, 1, 2, \ldots$, in view of (9.10). $\square$

**Corollary 12.** $(\gamma, \gamma; P) \cap (X, \gamma) = \phi$, *when* $X = c_0, c, \ell_\infty$.

The following results are also known [25].

**Theorem 104.** $A = (a_{nk}) \in (c_0, \ell)$ *if and only if*

$$\sup_N \sum_{k=0}^{\infty} \left| \sum_{n \in N} a_{nk} \right| < \infty, \tag{9.14}$$

*where $N$ is a subset of $\mathbb{N}_0 = \{0, 1, 2, \dots\}$.*

**Theorem 105.** $A = (a_{nk}) \in (\gamma, \ell)$ *if and only if*

$$\sup_{N,K} \left| \sum_{n \in N} \sum_{k \in K} (a_{nk} - a_{n,k-1}) \right| < \infty, \tag{9.15}$$

*where $N, K$ are subsets of $\mathbb{N}_0$. Further, $A \in (\gamma, \ell; P)$ if and only if (9.10) and (9.15) hold, where $(\gamma, \ell; P)$ is the subclass of $(\gamma, \ell)$ such that*

$$\sum_{n=0}^{\infty} (Ax)_n = \sum_{k=0}^{\infty} x_k, \ x = \{x_k\} \in \gamma.$$

We now have the following Steinhaus-type result.

**Theorem 106.** $(\gamma, \ell; P) \cap (c_0, \ell) = \phi$.

*Proof.* Let $A = (a_{nk}) \in (\gamma, \ell; P) \cap (c_0, \ell)$. Using (9.14) we have

$$\sum_{k=0}^{\infty} \left| \sum_{n=0}^{\infty} a_{nk} \right| < \infty,$$

which is a contradiction, since $\sum_{n=0}^{\infty} a_{nk} = 1$ for every $k = 0, 1, 2, \dots$, in view of (9.10), completing the proof. □

**Corollary 13.** $(\gamma, \ell; P) \cap (X, \ell) = \phi$, *when* $X = c_0, c, \ell_\infty$.

**Remark 11.** $(c, \gamma; P')$ *denotes the subclass of* $(c, \gamma)$ *such that*

$$\sum_{n=0}^{\infty} (Ax)_n = \lim_{k \to \infty} x_k, \ x = \{x_k\} \in c.$$

*In the context of Steinhaus-type theorems, we note that*

$$(c, \gamma; P') \cap (\ell_\infty, \gamma) \neq \phi.$$

## Exercises

1. Characterize the matrix class $(\ell_p, c)$, $p > 1$.

2. Characterize the matrix class $(\ell, \ell; P)$.

3. Find necessary and sufficient conditions for $A = (a_{nk}) \in (\gamma, \ell)$.

4. Find necessary and sufficient conditions for $A = (a_{nk}) \in (\ell, \gamma)$.

5. Characterize the matrix class $(\ell, \gamma; P)$.

6. Characterize the matrix class $(\gamma, \gamma; P)$.

7. Characterize the matrix classes $(\gamma, \gamma)$ and $(\gamma_\infty, \gamma)$.

8. Characterize the matrix class $(\gamma, c; P')$.

9. Characterize the matrix class $(c_0, \gamma)$.

10. Characterize the matrix class $(c_0, \ell)$.

11. Find characterizing conditions for $A = (a_{nk})$ to be in the classes $(\gamma, \ell)$ and $(\gamma, \ell; P)$.

# Bibliography


[1] G. Bachman and L. Narici. *Functional Analysis.* Academic Press, New York, 1966.

[2] S. Banach. *Théorie des opérations linéaires.* Chelsea, New York, 1955.

[3] R.G. Cooke. *Infinite Matrices and Sequence Spaces.* Macmillan, 1950.

[4] S.R. Foguel. On a theorem of A.E. Taylor. *Proc. Amer. Math. Soc.*, 9:325, 1958.

[5] J.A. Fridy. A note on absolute summability. *Proc. Amer. Math. Soc.*, 20:285–286, 1969.

[6] J.A. Fridy. Properties of absolute summability matrices. *Proc. Amer. Math. Soc.*, 24:583–585, 1970.

[7] C. Goffman and G. Pedrick. *First Course in Functional Analysis.* Prentice-Hall of India, New Delhi, 1974.

[8] G.H. Hardy. *Divergent Series.* Oxford, 1949.

[9] V.K. Krishnan. *Text Book of Functional Analysis.* Prentice-Hall of India, New Delhi, 2001.

[10] B.V. Limaye. *Functional Analysis.* Wiley Eastern Ltd., 1981.

[11] L.A. Lusternik and V.J. Sobolev. *Elements of Functional Analysis.* Gordon and Breach Science Publishers, 1961.

[12] I.J. Maddox. On theorems of Steinhaus type. *J. London Math. Soc.*, 42:239–244, 1967.

[13] I.J. Maddox. *Elements of Functional Analysis.* Cambridge University Press, 1977.

[14] P.N. Natarajan. A Steinhaus type theorem. *Proc. Amer. Math. Soc.*, 99:559–562, 1987.

[15] P.N. Natarajan. Some Steinhaus type theorems over valued fields. *Ann. Math. Blaise Pascal*, 3:183–188, 1996.

[16] P.N. Natarajan. A theorem of Steinhaus type. *J. Analysis*, 5:139–143, 1997.

[17] P.N. Natarajan. Some more Steinhaus type theorems over valued fields. *Ann. Math. Blaise Pascal*, 6:47–54, 1999.

[18] P.N. Natarajan. Some more Steinhaus type theorems over valued fields II. *Comm. Math. Analysis*, 5:1–7, 2008.

[19] P.N. Natarajan. On the $(M, \lambda_n)$ method of summability. *Analysis (München)*, 33:51–56, 2013.

[20] P.N. Natarajan. *Sequence spaces and summability over valued fields.* Taylor and Francis, 2019.

[21] A. Peyerimhoff. *Lectures on Summability.* Lecture Notes in Mathematics, Springer, 107, 1969.

[22] R.E. Powell and S.M. Shah. *Summability Theory and Applications.* Prentice-Hall of India, 1988.

[23] W. Rudin. *Principles of Mathematical Analysis.* McGraw-Hill, 1964.

[24] G.F. Simmons. *Topology and Modern Analysis.* McGraw-Hill, 1963.

[25] M. Stieglitz and H. Tietz. Matrixtransformationen von Folgenraümen Eine Ergebmisübersicht. *Math. Z.*, 154:1–16, 1977.

[26] A.E. Taylor. *Introduction to Functional Analysis.* Wiley, New York, 1958.

# Index